Interpreting Data

It may be now asked, to what purpose tends all this laborious buzzling, and groping? To know,

1. The number of the People?
2. How many Males, and Females?
3. How many Married, and single?
4. How many Teeming Women?
5. How many of every Septenary, or Decad of years in age?
6. How many Fighting Men?
7. How much London is, and by what steps it hath increased?
8. In what time the housing is replenished after a Plague?
9. What proportion die of each general and perticular Casualties?
10. What years are Fruitfull, and Mortal, and in what Spaces, and Intervals, they follow each other?
11. In what proportion Men neglect the Orders of the Church, and Sects have increased?
12. The disproportion of Parishes?
13. Why the Burials in London exceed the Christnings, when the contrary is visible in the Country?

To this I might answer in general by saying, that those, who cannot apprehend the reason of these Enquiries, are unfit to trouble themselves to ask them.

from *Natural and Political Observations Made upon the Bills of Mortality.*
John Graunt (1662)

ALAN J.B. ANDERSON
Senior Lecturer in Statistics
University of Aberdeen

A FIRST COURSE IN STATISTICS

London New York
CHAPMAN AND HALL

First published in 1989 by Chapman and Hall Ltd
11 New Fetter Lane, London EC4P 4EE
Published in the USA by Chapman and Hall
29 West 35th Street, New York 10001

Photosetting by Thomson Press (India) Ltd., New Delhi
Printed in Great Britain by J.W. Arrowsmith Ltd, Bristol

ISBN 0 412 29560 1 (hardback)
0 412 29570 9 (paperback)

British Library Cataloguing in Publication Data

Anderson, Alan J.B., 1941–
Interpreting data.
1. Statistical analysis
I. Title
519.5

ISBN 0 412 29560 1
ISBN 0 412 29570 9 (Pbk.)

Library of Congress Cataloging in Publication Data

Anderson, Alan J.B., 1941–
Interpreting data.

Includes index.
1. Statistics. I. Title.
QA276.12.A42 1988 001.4'22 88-16195
ISBN 0 412 29560 1
ISBN 0 412 29570 9 (Pbk.)

To my wife

Contents

Preface xiii
Acknowledgements xiv
Introduction xv

1. *Populations and samples* 1
   1.1 Populations 1
   1.2 Samples 2
   1.3 How to choose a (simple) random sample 4
   Summary 7
   Exercises 7

2. *Tabular and graphical summaries of data* 9
   2.1 Introduction 9
   2.2 Kinds of data 9
   2.3 Summarizing qualitative data 10
   2.4 Summarizing quantitative data 14
   2.5 Population distributions for continuous variates 20
   Summary 21
   Exercises 22

3. *Numerical summary of data – sample statistics* 33
   3.1 Measures of location 33
   3.2 Which measure of location? 35
   3.3 Variance and standard deviation 36
   3.4 Notation 37
   3.5 Grouped data 38
   3.6 Variability of sample means 41
   Summary 42
   Exercises 43

4. *Association between variables* 46
   4.1 Introduction 46
   4.2 Correlation 47
   4.3 Simple linear regression 49
   4.4 A real association? 56
   4.5 Cause and effect 59

Summary 59
Exercises 59

5. *Testing hypotheses* **65**

5.1 Introduction 65
5.2 Probability 65
5.3 Probability distributions 67
5.4 Testing a hypothesis 72
5.5 Problems of hypothesis testing 78
5.6 Estimation 80
Summary 81
Exercises 81

6. *Published statistics* **85**

6.1 The collection of data 85
6.2 UK published statistics 86
6.3 Reliability 90
6.4 Consistency 91
6.5 Official statistics in the developing countries 92
Summary 94
Exercises 94

7. *Delving in tables* **95**

7.1 Introduction 95
7.2 Tabular presentation 95
7.3 Writing about tables 97
Summary 98
Exercises 98

8. *Changes with time – index numbers* **103**

8.1 Introduction 103
8.2 Simple indices 103
8.3 Weighted indices 104
8.4 Comments on the Laspeyres (L) and Paasche (P) indices 106
8.5 Changing the base 106
8.6 UK Index of Retail Prices 107
8.7 UK Index of Industrial Production 108
Summary 109
Exercises 109

9. *Demography – introduction* **111**

9.1 The importance of demographic studies 111
9.2 Dynamics of population change 112

9.3 Demographic transition 112
9.4 Age–sex profiles 113
9.5 Cohorts 119
Summary 120
Exercises 120

10. *Components of population change* **122**
10.1 Death 122
10.2 Marriage 130
10.3 Fertility (live births) 131
10.4 Migration 133
Summary 134
Exercises 134

11. *Population projection* **139**
11.1 Reliability of projections 139
11.2 An example 140
Summary 142
Exercises 142

12. *Changes with time – time series* **144**
12.1 Introduction 144
12.2 Smoothing times series 144
12.3 Components of time series 147
Summary 150
Exercises 150

13. *Data analysis and the computer* **152**
13.1 The computer 152
13.2 The analysis 153
Summary 160
Exercises 160

14. *Social surveys* **161**
14.1 The basic framework 161
14.2 Pilot study 161
14.3 Methods of data collection 162
14.4 Confidentiality 165
14.5 Questionnaire design 165
14.6 Analysis 170
14.7 The report 171
Summary 171
Exercises 172

15.	***Schemes of investigation***	**174**
	15.1 Introduction	174
	15.2 Problems	174
	15.3 Matching	179
	15.4 Randomization	180
	Summary	181
	Exercises	181
16.	***Controlled experiments***	**183**
	16.1 Twin studies	183
	16.2 'Blocking'	184
	16.3 Confounding	186
	16.4 Factorial treatment combinations	186
	16.5 Clinical trials	188
	Summary	189
	Exercises	189
17.	***Sampling schemes***	**191**
	17.1 Introduction	191
	17.2 Systematic sampling	191
	17.3 Simple random sampling from a sequence	192
	17.4 Stratified sampling	193
	17.5 Multistage sampling	195
	17.6 Quota sampling	198
	17.7 Standard errors	198
	Summary	199
	Exercises	199
18.	***Longitudinal studies and interrupted time series***	**201**
	18.1 Prevalence	201
	18.2 Incidence	201
	18.3 Cross-sectional versus longitudinal surveys	202
	18.4 Interrupted time series studies	203
	Summary	208
	Exercises	209
19.	***Smoking and lung cancer***	**210**
	19.1 Introduction	210
	19.2 Experiments	210
	19.3 Published figures	210
	19.4 Retrospective studies	211
	19.5 Prospective studies	211

Summary 212
Exercises 212

20. *An overview* **214**

Glossary of symbols **217**
Index **219**

Preface

In his Presidential address to the Royal Statistical Society in 1974, Professor D.J. Finney summarized the two primary duties of a statistician as

1. '...to interpret quantitative information validly and usefully and to express himself in terms intelligible beyond the confines of statistics.'
2. '...to advise on the acquisition and collection of data...the design of an experiment, the rules by which a sample is selected, the manner in which records are prepared for statistical analysis,...all that relates to recording, to the planning of questionnaires, to the choice and exact definition of measurements, to copying and checking.'

These two aspects of statistics are essentially inseparable and, though each chapter in this text appears to focus on one or the other, the reader should see the topics covered as offering merely different perspectives of a unified whole. The order of chapters is a little arbitrary and many could be omitted by the reader involved in a specific application domain. The biologist, for instance, might relegate Chapters 6, 7, 8, 14, and 18 to a secondary level of interest; the social scientist might correspondingly choose to neglect Chapter 16. Chapter 5 is a highly-compressed introduction to statistical inference, material that usually occupies the whole of an introductory course but which, in this book, is down-weighted to give greater prominence to more fundamental principles of interpretation; this chapter could therefore be ignored in a first reading. Nevertheless, it is likely that the text will be helpful when studied in conjunction with a traditional methods-oriented statistics course.

The presentation assumes very little mathematical knowledge though, of course, data analysis cannot be free of arithmetic manipulation. As well as making the contents available to as wide a readership as possible, the absence of theoretical derivation serves to emphasize the fact that the concepts of statistics are independent of the mathematical and computational tools that may be used in their development.

The exercises at the end of each chapter cover a broad spectrum of application areas. In the earlier chapters these problems are mainly abstracted from research publications to help make the ideas concrete. The later exercises, particularly those relating to methods of data collection, are often susceptible to different approaches, thus reflecting the vagaries of real life research planning. Some of the exercises introduce material not covered in the main text and the aim of all of them is to stimulate thought rather than to elicit cook-book solutions.

Acknowledgements

This book has evolved from material used in a first-year undergraduate statistics course at the University of Aberdeen for the past twelve years and I am thus grateful to my colleagues and many students for critical comments. Dr Frank Yates, the mentor of my early professional years, kindly read a draft of the text and offered many constructive suggestions though, of course, any errors or omissions that remain are my responsibility. Likewise the comments of my wife, Bridget, on the content and presentation have been most helpful. I am indebted to the publishers of the Sunday Times and the Aberdeen Press and Journal for permission to publish extracts and to Her Majesty's Stationery Office for permission to reproduce a number of tables and figures. Mrs Lilian Reid successfully persuaded the computer to produce all the graphical material and I express my thanks to her and to Mrs Stella McHardy and Mrs Gail Willox who prepared the manuscript with enthusiasm and efficiency.

Introduction

When we look at nature, it has much of the appearance of a wobbling jelly.

In the physical sciences such as physics or chemistry, the wobble is due to errors of measurement that are so small that it is usually easy to discern underlying 'laws'. Thus, if we hold gas at a constant temperature and increase the pressure on it, the volume decreases proportionately. Boyle's experiment of 1662 has been repeated so often in school laboratories that there can be no doubt about the truth of his law, observed discrepancies being readily explained by inability to maintain the gas at the prescribed temperature without fluctuation or to make exact determinations of pressure and volume.

In the biological sciences and medicine, on the other hand, natural variation takes over as the main cause of wobble and we can make statements only about the average behaviour of plants, animals and humans. Such unpredictability is a fundamental property of living organisms. In the study of human societies, this variation is greatly magnified and compounded with changes through time, so that the shape of the jelly is even harder to perceive.

The word 'statistics' has the same root as 'state', 'stationary', 'static', etc., and the science (or is it the art?) of statistics is concerned in trying to freeze the wobbling picture that nature shows us. The term first appeared in mid-eighteenth century political and sociological studies but Darwin's theory of evolution stimulated a rapid development of statistical techniques in biological and agricultural research and, nowadays, unified methods find application in industrial production, epidemiology, literary criticism, educational assessment, financial planning, drug trials, actuarial prediction, etc. In quick succession, the practising statistician may help unravel the walking habits of crabs, corrosion rates in oil pipelines, the emotional involvement of psychotherapists with their patients, the tendency of children to occupy the same niche in society as their parents, the identification of suspects by witnesses, the estimation of volume of timber in a forest, employers' attitudes towards alcoholic staff members, and so on. These lists are given, not so much to inspire readers to become tomorrow's professional statisticians, as to underline the need for people in many disciplines to be able to communicate about statistics. Unfortunately, a good deal of jargon is involved, but common sense is much more important as a prerequisite than mathematics.

In most statistics courses, the main emphasis is placed on the natural variation component of the jelly's wobble; removal of this component is central to the making of inferences about nature when we can look at only a part of it. This book is concerned principally with all the other kinds of uncertainty that can arise in the course of research and decision making. The text relates partly to the collection of data to answer specific questions and partly to the interpretation of these data and data collected by others.

1
Populations and samples

1.1 POPULATIONS

If statistical methods are to be of use in answering a particular question, one of the first steps must be specification of the so-called **universe of discourse**, that is to say, the collection of actual or hypothetical items defined by some common characteristic to which the question and answer are to apply. In statistical terms, these items are called **units** and the aggregate of all the units is termed the **population**. These units can indeed form the human population of a country or region or town; but equally the word could be applied to the collection of all primary schools in Scotland in a study on education, to all first-division football matches in the last five years in a study of hooliganism, to all Concorde take-offs from a certain airport in a study of aircraft noise, and so on. With care, it is possible to talk about a population of measurements but, for example, the population of trees in a wood and the population of their heights are, strictly speaking two different populations (of the same size). We should be wary of blurring this kind of distinction.

Sometimes populations can be enumerated, that is to say, the individual units can be identified and listed. Often this is not feasible, perhaps because the notional population changes too rapidly over time; the population of homeless in a large city, the badger population of England, the population of diseased trees in any forest are examples of reasonably well-defined populations that could not be numbered. Indeed, particularly in the case of animal studies, the sole objective can be to determine the size of the population at a given point in time.

Frequently – perhaps even usually – the population under consideration is conceptually infinite in size because additional units can always be found or generated experimentally. Consider, for example, tosses of a coin, car crashes at 60 mph, letters sent from town A to town B, attacks of malaria. The idealization of some aspect of nature as an unlimited number of imaginary units is an important idea even though we recognize that the world's finite resources of energy and time could never allow attainment of such a population. This is the context of statements like 'This drug has a greater effect than that drug' or 'This make of car seat belt is better than that'. It is easy to see how dangerous these assertions could be if the hypothetical population for which they apply has not been well defined.

Clearly, the examination of every unit of a population (a **census**) can be impossible or impracticable or simply inefficient and we must turn to the process of sampling.

1.2 SAMPLES

Sampling is the selection of some units of a population to represent the whole aggregate and hence we must make certain that the sample members are as typical as possible in relation to the objectives of the study. Before considering how to achieve this, let us look at some general advantages and disadvantages of sampling.

1.2.1 Advantages of sampling

Sampling has several potential benefits even when a complete census might be feasible:

1. If we can assume that the population is well mixed, then it need not be enumerated. We must effectively make such an assumption with all infinite populations. We can readily accept, for example, that a blood sample taken from the arm will produce a selection of blood cells that are representative of all those in the body. But use of a single handful of grain from the hopper of a combine harvester to assess protein content for a whole field is clearly much more debatable.
2. Quality control in industrial production usually demands a sampling approach. This may be because the required tests are destructive in some sense. For example, since one cannot check the lifetimes of batteries without rendering them unsaleable, only a small proportion of output can be examined. Alternatively, the time taken to inspect a product (nuts and bolts, for instance) can simply be so great as to make the process uneconomic if every item produced were to be checked.
3. Sampling is generally cheaper (though the cost per unit studied will usually be greater than for a census). The available funds may therefore permit the use of better methods of data collection and validation, more experienced clerical staff, and so on, so that the quality of information collected is higher. Thus it is often possible to obtain more information per unit because the finite resources of time, money and manpower permit a more extensive investigation of the reduced number of units. (Unfortunately researchers on constrained budgets tend mistakenly to devote their funds to a large sample at the expense of all other aspects of the study.)
4. If the study requires participation by members of the general public (for example, in answering a questionnaire), it may be possible to collect more information per person than by a census simply because, for psychological reasons, people are more cooperative when they feel themselves specially chosen.
5. Data from a sample are generally available for analysis more rapidly and the analysis itself can be completed more quickly so that the findings are less likely to be out of date. Improvements in data processing techniques are steadily reducing this difference but, for instance, summaries of the United Kingdom

census of 1971 were not complete until 1977 for a questionnaire that had been devised 10 years previously.

1.2.2 Problems of sampling

The main disadvantage of using a sample is that we have to tolerate a greater degree of uncertainty. This uncertainty arises from two causes:

1. The **natural variation** due to chance differences among members of the population that will be reflected in variability over the possible samples. This component of uncertainty can be evaluated by statistical methods that are discussed only superficially in this book (in Chapter 5).
2. Inadequate definition of the population (which, naturally, cannot occur with a census) or shortcomings in the method of choosing the sample. Errors of this kind give rise to the component of uncertainty called **bias.** Such errors are:
   (a) Subjective choice of sample: experiments have shown that individuals, asked to make a selection of typical stones from a large collection of flints, will tend to pick those that are larger than average; observers, asked to select young wheat shoots 'at random' also choose those that are, on average, too tall.
   (b) Sampling from an inadequate list: thus, a sample chosen from the telephone directory obviously cannot provide a good estimate of, say, the average income of the whole population. A famous illustration is the disastrous attempt to predict the results of the United States presidential election of 1936 by such means; nowadays, however, telephone sampling is becoming more acceptable.
   (c) Incomplete response in postal surveys: for example, those that take the trouble to return questionnaires may be particularly interested in the objectives of the study and hence are not typical of the whole population.
   (d) Substitution in the sample: in a survey involving house visits, for instance, the interviewer may take the next-door household whenever there is no reply; this 'cheating' will result in a preponderance of houses that are often occupied (by families rather than single people or by elderly people). Bias could also occur in a botanical survey of the plants growing in metre-square sample quadrats of a moorland if those wholly occupied by rocks or water were replaced by adjacent areas.
   (e) 'Edge effects': plants growing at the perimeter of a plot in an agricultural experiment will be unrepresentative and are normally discarded. Likewise, student attendances at the start of term cannot be taken as typical of the whole year; and the pulse of a patient immediately after lying down will not characterize the true supine heartbeat pattern.

A characteristic of bias is that it usually cannot be reduced by increasing the sample size since every unit sampled is likely to misrepresent in the same way. If a grain-weighing machine is consistently over-recording by 50 g, the average yield per square-metre plot will tend to be biased upwards by 50 g no matter how many such plots are sampled. Notice, however, that the difference in yield between, say, two varieties of grain will be correctly estimated since, if the same machine is used for all determinations, the biases will cancel out.

Bias can be eliminated by

1. ensuring that the required population is indeed the one sampled (though this may be very difficult or indeed impossible);
2. choosing a **random** sample, that is to say, one in which every member of the sampled population is guaranteed an equal chance of being included in the sample.

1.3 HOW TO CHOOSE A (SIMPLE) RANDOM SAMPLE

It is usual to indicate the number of units in the population by N and the number of units in a particular sample by n. To select a random sample of size n, we need

1. a list of all N units of the population, numbered 1 to N
2. some mechanism for selecting n *different* numbers from the range 1 to N.

(Samples for which a unit can appear more than once – sampling *with replacement* – are sometimes useful but we shall ignore this here.) Each of the N numbers must have an equal chance of selection, i.e. the mechanism must be equivalent to drawing from a hat, spinning a top with N edges, etc. As a result, every possible sample of size n will have an equal chance of being the one chosen.

Computers can produce the sample numbers for us; an obvious example is ERNIE, the machine that selects the winners of the United Kingdom Premium Bond scheme. But, if n is not too large, tables of previously generated pseudo-random digits (random number tables) provide an easy alternative that is also instructive.

Suppose we want a sample of $n = 10$ from a population of $N = 100$. Consider the random digits given in Table 1.1. (The gaps in the table are sometimes visually convenient or can be ignored.) We shall need two-digit numbers, using '00' to represent '100'.

Let us assume for the present that we have somehow fixed our starting point at the tenth row, fifth digit, i.e. 83 is our first number. Reading across is quite acceptable but downwards is possibly a little easier and we shall do this

Table 1.1 Random Numbers

25 85	52 40	80 50	80 78	58 42	11 31	85 77	77 25	16 08	54 37
58 73	38 58	78 92	12 38	43 41	31 77	97 30	33 45	00 17	60 35
66 04	44 17	00 38	61 37	54 84	38 54	05 96	18 96	20 83	65 29
96 22	27 19	23 83	09 18	22 67	17 31	63 08	80 18	68 08	47 88
29 70	86 38	78 04	51 58	31 92	12 30	98 81	53 09	07 63	23 16

Table 1.1 (*Contd.*)

83 86	48 37	00 91	51 91	62 88	04 62	94 63	12 46	51 12	55 22
24 10	43 44	80 33	91 59	46 71	46 72	33 99	13 16	51 34	84 49
26 45	22 19	59 42	70 02	50 23	78 14	24 19	86 33	37 11	65 36
59 58	24 97	89 51	48 61	85 92	96 83	36 62	06 90	17 23	92 17
14 19	83 76	52 64	91 14	70 66	74 05	47 54	26 81	32 14	91 27
27 70	78 05	51 37	28 77	39 03	63 18	66 74	90 45	23 47	49 80
98 13	78 10	97 38	27 61	27 91	81 70	55 28	35 37	45 76	48 82
94 08	72 66	37 34	44 45	32 26	04 37	80 55	29 88	65 99	63 64
49 76	05 70	91 47	21 85	45 96	57 06	08 49	03 91	19 89	87 10
17 55	25 70	35 57	96 81	90 91	91 05	56 31	55 22	15 63	67 57
57 83	92 42	51 67	49 06	19 44	23 43	39 33	65 52	69 76	98 66
54 36	96 45	91 71	43 43	93 04	71 90	40 81	82 26	56 84	31 32
95 36	83 47	04 07	56 30	59 82	14 82	28 69	27 22	19 47	18 08
61 12	39 28	72 79	97 30	82 43	58 53	40 49	99 71	09 76	60 35
26 27	83 09	88 15	45 34	19 86	67 13	18 50	45 33	02 48	34 58
45 24	50 85	51 89	13 09	85 23	33 26	92 73	47 94	81 86	25 94
44 77	24 50	96 30	18 51	79 05	26 81	87 43	50 71	99 55	32 09
19 79	28 20	23 65	41 96	95 76	81 06	28 11	36 66	67 16	33 42
55 17	44 90	96 22	30 63	61 12	03 95	61 40	29 69	40 05	50 67
73 57	60 29	70 47	76 17	78 45	18 43	68 24	02 39	16 40	86 60
89 02	98 00	17 62	52 94	92 56	80 00	00 26	31 95	24 34	84 41
82 02	00 81	89 86	31 61	31 23	38 70	16 15	93 06	97 44	33 63
52 01	93 94	50 55	19 39	99 45	33 41	40 40	71 54	95 11	19 13
64 80	82 89	45 39	32 76	89 53	80 65	64 77	23 81	89 46	93 31
78 83	74 73	25 33	78 96	68 49	40 11	35 36	17 50	41 41	48 02
16 71	31 07	82 88	89 96	44 21	88 20	22 77	23 29	83 14	38 06
98 90	44 87	42 03	23 23	87 14	30 08	25 17	46 51	03 58	46 62
34 95	90 75	63 46	67 53	13 59	91 93	58 51	43 69	78 88	37 65
30 27	22 59	07 85	95 97	61 29	31 56	45 21	26 71	99 18	31 37
00 08	48 54	49 72	89 48	59 49	15 41	70 48	69 97	49 77	29 22
33 57	45 38	37 68	28 69	03 29	50 80	59 48	61 60	85 08	65 40
54 34	43 44	14 00	94 50	67 57	68 15	08 20	25 55	19 37	75 38
74 21	68 80	20 75	39 51	58 77	39 47	93 74	54 29	26 18	18 59
65 09	82 46	75 80	58 91	62 39	49 97	01 57	90 83	42 44	47 34
10 43	48 61	60 65	96 45	17 33	03 50	78 26	85 72	61 01	67 75
01 94	95 31	34 40	15 77	57 25	94 76	77 95	34 61	57 80	84 91
57 72	84 03	52 33	45 45	54 58	76 24	11 70	47 64	31 34	00 52
85 89	47 95	04 23	30 42	91 63	50 27	31 88	37 53	93 46	14 34
11 43	25 72	26 18	52 81	68 53	65 65	62 77	29 63	81 17	21 73
73 62	52 21	84 78	63 17	45 28	35 53	60 79	80 43	87 65	80 44
55 91	18 37	95 71	45 13	16 10	65 76	74 52	87 59	10 93	00 48
26 40	72 27	23 11	80 68	17 82	54 90	09 34	15 03	92 54	62 71
28 53	58 92	62 13	56 50	36 18	96 48	96 31	03 67	94 99	60 37
39 70	30 58	87 42	52 62	06 97	20 58	52 63	22 29	13 57	34 10
75 78	85 97	01 92	55 74	85 83	02 12	07 93	32 86	31 21	41 32

so that we get

83 78 78 ...

The second 78 is no use to us so we must discard it and continue until we have ten different unit numbers, i.e.

83 78 78 72 05 25 92 96 83 39 83 50 24

These, then, define our sample.

Easy! But what if we had wanted a sample from a population of size $N = 153$? We would obviously have to use three columns; and (starting, for example, from the same point) we would generate many useless numbers

837 780 781 726 057 ...

Thus, 57 is apparently the first that we can use; $(1000-153)/1000 = 84.7\%$ of selections will, on average, be outside the range 1 to 153 and will therefore be 'wasted'.

We can remedy the situation somewhat as follows. Consider 201 to 400 as corresponding to 1 to 200, the same with 401 to 600, and so on ('1000' now being represented by '000'). Each number in the range 1 to 153 has now five ways of occurring at each selection but all the numbers remain equally likely and now only $(200-153)/200 = 23.5\%$ on average of selections are wasted. Thus, the sample will be

837	780	781	726	057	257	924	964
≡ 37	≡ 180	≡ 181	≡ 126	≡ 57	≡ 57	≡ 124	≡ 164
	no use	no use			repeat		no use

834	392	830	508	245	282	449
≡ 34	≡ 192	≡ 30	≡ 108	≡ 45	≡ 82	≡ 49
	no use					

We could obviously devise a system that would waste even fewer selections but the greater arithmetic complexity would result in little time being saved. Whatever method is used, it is essential to check that the 'equally likely' rule has been followed.

It remains only to indicate how the 'random' starting point in the table should be chosen.

1. We could close our eyes and attack the page with the proverbial pin. Of course, this is not as random as it should be. (We would tend to land in the middle of the table and run a considerable risk of producing overlapping samples on different occasions.) But we could use the digits we find there to specify the row and column of the starting point that we shall actually use.

With 50 rows and 40 columns, we shall require two-digit numbers for each, treating 51 to 100 as equivalent to 1 to 50 (and, in the case of the columns, omitting 41 to 50 and 91 to 100). Suppose our pin lands on the 36th row and 31st column; this gives the two-digit 'random' numbers 60 and (below) 55 corresponding to a starting point at row 10 and digit 5 as was used above.

2. An easier procedure, if the same table is always used, is simply to carry on from immediately after the sequence obtained in the last use of the table.

All good sampling procedures are built around the idea of the simple random sample. Use of the adjective 'simple' implies that, in practice, more complicated prescriptions are often needed and we shall return to this subject in Chapter 17.

SUMMARY

The aggregate of units in the universe of discourse of a study defines the (finite or infinite) population to be considered. A sample that is truly random can be representative of the population from which it is drawn. Sampling has several advantages and disadvantages over the taking of a complete census. The main interpretive problems arise through the uncertainties inherent in sampling variation and bias distortions due to careless planning and operation of the study.

EXERCISES

1. (In many ways, this question is just a test of common sense; however, the style of thinking needed is fundamental to the interpretation of data.)

   Examine each of the following statements and the conclusion drawn from it. Assuming that the statement is true, present carefully reasoned arguments against the validity of the conclusion.

   (a) The average income of men in a particular country is £10 000 per annum, and of women is £8 000 per annum. Hence the average married couple earn £18 000 between them.
   (b) Of men sentenced to life imprisonment, only 10% are ever convicted of another offence whereas, of men convicted to six months or less, 80% are eventually reconvicted. Hence longer sentences are a much more effective deterrent.
   (c) Many insurance companies offer lower premiums to women drivers than men. This is evidence that women are better drivers.
   (d) Since the introduction of comprehensive schools, standards of reading and writing among school children have declined. This proves that the change from selective schools was a serious mistake.
   (e) Among all methods of transport the number of deaths per passenger mile is lowest for journeys by spacecraft. Hence this is the safest way to travel.

2. Thinking in terms of well-defined populations and useful measurements on feasible samples how would you attempt to answer the following questions?

   (a) How successful is the Samaritans organization in preventing suicide?

(b) Is the potato variety King Edward more resistant to disease than the variety Arran Pilot?
(c) Are the children of single-parent families backward in reading ability at the age of 7 years?
(d) To what extent is the sex of a crocodile determined by the incubation temperature experienced by the egg from which it hatched?

3.

Last year a random
selection of 26
RARE STAMPS
showed a 73.87%
increase in value.

Criticize this advertisement placed by a stamp dealer.

4. It is often necessary to select random points in an area of ground. This may be, for example, to define the positions of small sample plots (**quadrats**) within which plant species are to be recorded or to select random trees in a forest – the nearest ones to the points chosen.

A suggested method is to stand at the corner of the area, take a random compass bearing and then a random number of paces in that direction towards the far boundary. What is wrong with this?

(**Hint:** Draw a picture; would you expect approximately equal numbers of points within ten paces of the start and between ten and twenty paces and so on?)

Think of an alternative method; how practicable is it?

5. A rectangular forestry plantation has trees laid out in 176 rows each containing 80 trees. A suggested sampling mechanism involves randomly selecting a row and then randomly choosing a tree in a row.
(a) Use this method to produce a random sample of 10 trees.
(b) If the row lengths were irregular, varying from (say) 40 to 120 trees, explain why the above method would not produce a random sample of trees.

6. (a) Make a list of all the 15 possible samples of size 2 that can be drawn from a population of 6 members labelled A, B, C, D, E, F.
(b) For random sampling, explain intuitively why each of the 15 samples is equally likely to be the one chosen.

2
Tabular and graphical summaries of data

2.1 INTRODUCTION

One of the founders of modern statistics, R.A. Fisher, presented the subject matter of statistical science in three interlocking aspects:

1. the study of **populations** – we have introduced this fundamental idea in Chapter 1
2. arising from 1, the study of **sampling variation** – we have hinted at this component and will return to it in Chapter 5 though some of the underlying ideas are introduced in this chapter and the next
3. the study of methods for the **reduction of data** – that is to say, the means by which the relevant information in a large collection of observations can be extracted and presented.

Much of practical statistics is concerned with this process of summarizing and describing data and this chapter and the next two outline some of the basic principles. The ideas are mostly presented in terms of the sample, but many apply equally to the population which can be regarded as the special sample where $n = N$ (a census).

Data consist – notice that 'data' is a plural noun – of observations of one or more attributes on each of a number (n) of individual units. These attributes are usually called **random variables** or **variates.** Thus, 'country represented' by athletes in an international competition, 'percentage dry matter' of yields in a cereal crop experiment and 'period of loan' in a library's study of book lending are all examples of random variables.

2.2 KINDS OF DATA

Random variables can be of two main types:

1. **Qualitative.** Here, the attribute is such that there is no sense of ordering involved. Thus 'red hair' is not bigger, better or greater than 'black hair' nor vice versa; likewise, 'maleness' is neither superior nor inferior to 'femaleness'. Some people use the term **categorical** to describe this kind of variate. Normally such categories do not overlap so that the attribute defines mutually exclusive subpopulations.

2. **Quantitative.** In this case, the observations have a very definite sense of ordering; there are *more* students in a class of 50 than in one of 40, a height of 181 cm is *taller* than a height of 180 cm, 16 February 1976 came *before* 22 March 1977, an election swing of 4.3% is *greater* than one of 3.4%, to come second in a race indicates a *better* performance than coming sixth. In fact, these examples reveal three kinds of quantitative data:

   (a) **Discrete** data: here, we are talking about counts that can take only integer values 0, 1, 2, 3,... and not fractions.
   (b) **Continuous** data: here, it is possible, with sufficiently precise instruments, to observe any value within the range of feasible values of the attribute. Thus, if we could measure only in centimetres, two men might both appear to be 180 cm tall; but, if we could measure in millimetres, we might find heights of 180.2 cm and 180.3 cm; we would need to go to tenths of a millimetre to distinguish men 180.23 cm and 180.24 cm tall; and so on. Height is therefore an attribute that can have any of an infinity of possible ordered values between the feasible extremes. (Scores such as IQ, examination marks expressed as percentages, etc., can have only integer values, but are nevertheless classed as continuous since the underlying attribute is continuous.)
   (c) **Ordinal** data: here, the observations are of the kind 'third top' or 'fifth from the end'. We shall not be very much concerned with such data.

2.3 SUMMARIZING QUALITATIVE DATA

Almost the only thing one can reasonably do with qualitative data is to count the numbers of items in different categories. For small samples, this can be done using some form of simple tally system.

Example 2.1
Suppose the blood groups of a sample of 25 males are:

A B A A A AB O A A A O B O B A B O B O A B B A A O

The four categories are A, B, AB, and O. We go through the 25 values counting by means of 'five-bar gates':

Category	Tally	Count
A	𝍸 𝍸 |	11
B	𝍸 ||	7
AB	|	1
O	𝍸 |	6
		25

For larger amounts of data, we would use a computer. Either way, the end product is a frequency table.

2.3.1 Frequency tables

Example 2.2

Frequency table of intended degree for BSc students at Aberdeen University 1984/85

BSc (Agriculture)	155
BSc (Engineering)	334
BSc (Forestry)	100
BSc (Pure Science)	1271
Total	1860

Source: University of Aberdeen *Annual Report* 1985

Notice that any summary table must have a descriptive title (normally preceding the table) and, if at all possible, an indication of the source of the information. Notice also that, because we do not wish to imply any ordering, the types of degree are listed in alphabetical order.

We might have information permitting a further breakdown of the data to give a **two-way** frequency table of degree by sex, as follows.

Male/female breakdown of BSc students at Aberdeen University 1984/85

	Men	Women	Total
BSc (Agriculture)	103	48	155
BSc (Engineering)	304	30	334
BSc (Forestry)	93	7	100
BSc (Pure Science)	636	635	1271
Total	1140	720	1860

Source: University of Aberdeen *Annual Report* 1985

Notice how the so-called **margins** of this table (both labelled 'Total') give the one-way tables for type of degree and sex. Obviously one can produce more complicated tables, but few people like looking at anything more than three-way breakdowns.

Sometimes it is useful to express the counts in a table as percentages (usually to one decimal place) of the overall total. This helps comparison across tables (relating, to, for example, different years or different universities) provided the overall totals do not differ too greatly. Percentages of a margin can likewise be constructed, as follows.

Male and female percentages for BSc students at Aberdeen University 1984/85 categorized by type of degree

	Men	Women
BSc (Agriculture)	69.0	31.0
BSc (Engineering)	91.0	9.0
BSc (Forestry)	93.0	7.0
BSc (Pure Science)	50.0	50.0
Total	61.3	38.7

Source: University of Aberdeen *Annual Report* 1985

Because the row totals must now be 100.0, the right-hand margin is omitted. Since we now do not know from the table the number of students involved, the figure (1860) should be stated somewhere nearby and possibly also the totals for each type of degree, so that the original figures can be reconstructed.

2.3.2 Graphical summaries

Often it is helpful to present results graphically. Any such diagrams must have titles and any necessary explanatory annotation of axes, etc., and, if possible, an indication of the source of the data.

Two methods of presentation are commonly used for qualitative information presented in a frequency table, but both can be applied whenever we have proportions or percentages, e.g. proportions of agricultural land in a county allocated to rough grazing, temporary grazing, permanent grazing, cereals, non-cereals, etc.

(a) Bar diagrams

Such diagrams can misrepresent the truth through bad choice of scale for the vertical axis, so conventionally (and this applies to all graphs) the vertical axis should be such that the height of the diagram is between three-quarters and one-and-a-half times the width. Figure 2.1 shows a bar diagram for science students at Aberdeen.

(b) Pie diagrams (π charts)

These are like pies – the numbers of degrees in the slices being proportional to the

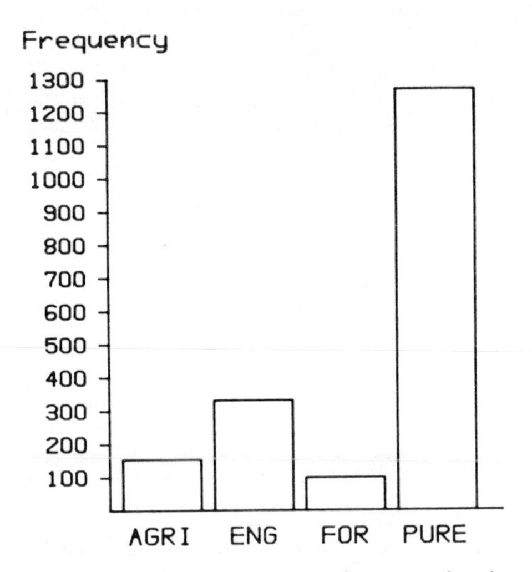

Fig. 2.1 Bar diagram of types of degree for BSc students at Aberdeen University 1984/85

frequencies, i.e.

$$\frac{x^\circ}{360^\circ} = \frac{\text{count in category}}{n}$$

i.e. $x = 360 \times$ proportion in category.

Hence for our Aberdeen University science undergraduates in Example 2.2 the pie diagram of Fig. 2.2 can be constructed as follows:

$$\text{Agri:} \quad 360 \times \frac{155}{1860} = 30.0^\circ$$

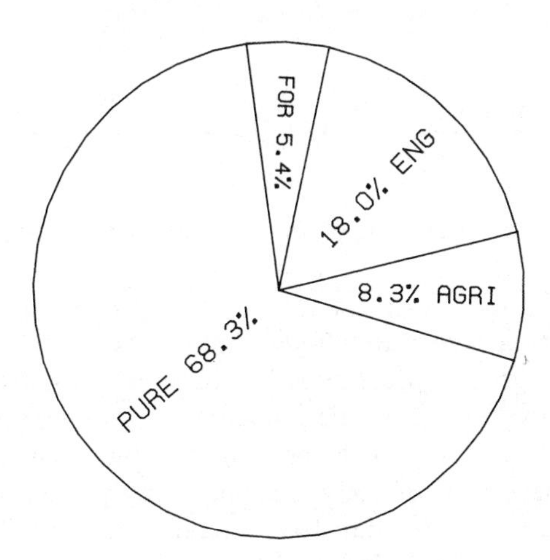

Fig. 2.2 Pie diagram of types of degree for BSc students at Aberdeen University 1984/85

$$\text{Eng:} \quad 360 \times \frac{334}{1860} = 64.6^\circ$$

$$\text{For:} \quad 360 \times \frac{100}{1860} = 19.4^\circ$$

$$\text{Pure:} \quad 360 \times \frac{1271}{1860} = 246.0^\circ$$

check total 360.0°

2.4 SUMMARIZING QUANTITATIVE DATA

How can we present graphical or tabular summaries of quantitative data?

With discrete data having only a moderate number of possible values, say, fewer than 30, no great problem arises. We can use a tally chart to construct a frequency table for the number of times each possible value occurs in the data.

Example 2.3

Numbers of children of 20 part-time teachers aged 40.

Data: 0 2 2 1 2 1 0 2 4 0 2 1 2 3 3 0 3 1 1 2

No. of children	Tally	Frequency
0	||||	4
1	𝍸	5
2	𝍸 ||	7
3	|||	3
4	|	1
		20

We could present these frequencies in a variant of the bar diagram called a **histogram** of which Fig. 2.3 is an example.

For continuous data, where an infinite number of different values are possible, we must obviously group values together in some way. (In fact this is done automatically to some extent by the crudeness of the measuring instrument or scoring system. Thus, if we are measuring a subject's height to the nearest centimetre, an observation of 160 cm means simply 'some value equal to or greater than 159.5 cm and less than 160.5 cm'; likewise, an IQ of 107 indicates a range from 106.5 up to but excluding 107.5.)

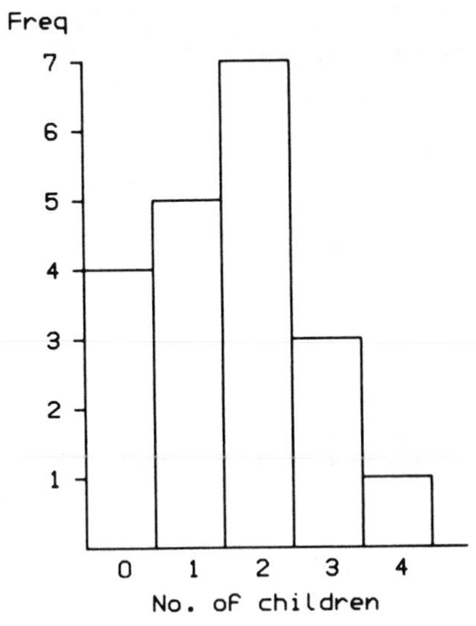

Fig. 2.3 Histogram of numbers of children of 40-year-old part-time teachers

Example 2.4
In an experiment on petrol consumption, 25 cars of the same make were driven steadily until each had used 1 gallon of the same brand of petrol. The distances travelled (in miles) were:

33.2	35.1	31.4	30.6	32.7
36.2	35.9	34.9	33.2	31.8
33.3	34.7	31.8	30.9	32.4
33.1	34.2	32.8	31.6	31.8
31.2	33.1	33.7	34.1	34.0

Clearly there are at least 57 possible values

$$30.6, 30.7, \ldots\ldots 36.2.$$

Let us reduce the measurements to crude miles per gallon. There is always an arbitrary element in this formation of classes relating to

1. width of classes – often decided by the need to have from about 5 to 30 classes
2. position of the class mid-points – often settled by some long-standing convention or by making the mid-points coincident with integral values.

By convention, the lower limit but not the upper is included in a class. We could therefore choose to have classes '30 and over up to 31' or '29.5 and over up to

30.5', etc. – no other scheme seems sensible. We shall arbitrarily choose the latter. Figure 2.4 shows the histogram of petrol consumption.

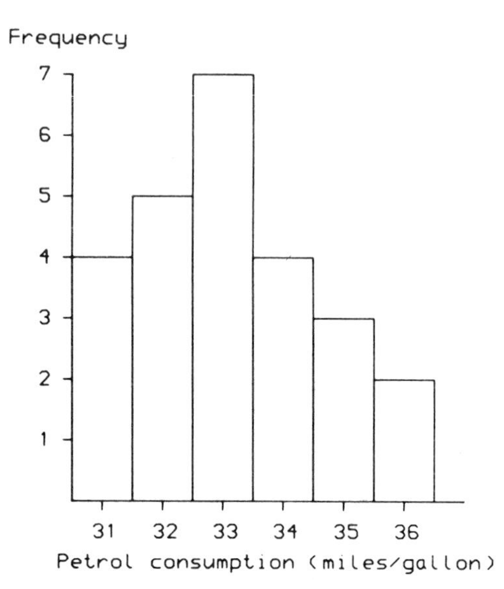

Fig. 2.4 Histogram of petrol consumption in an imaginary experiment

Hence

Miles/gallon	Tally	Count
30.5–31.4	||||	4
31.5–32.4	~~||||~~	5
32.5–33.4	~~||||~~ ||	7
33.5–34.4	||||	4
34.5–35.4	|||	3
35.5–36.4	||	2
		25

So far, we have used only equal class widths so that it has not mattered whether we thought of the frequencies as being represented by the heights or the areas of the 'boxes'. But it often makes more sense to have *unequal* class widths and, from now on, we must be clear that, *in a histogram, the frequencies are proportional to the areas of the boxes.*

Example 2.5

Frequency table of number of days absence from school in a year for a sample of 111 children

No. of days absent	Frequency
0	11
1	17
2	13
3– 5	15
6– 8	24
9–11	6
12–14	3
15–25	22
	111

Since the 3–5 category has a width of 3, the box must have a height such that

$$3 \times \text{height} = 15 \qquad \text{i.e. height} = 5 \qquad \text{and so on.}$$

Figure 2.5 shows the histogram of days absent from school.

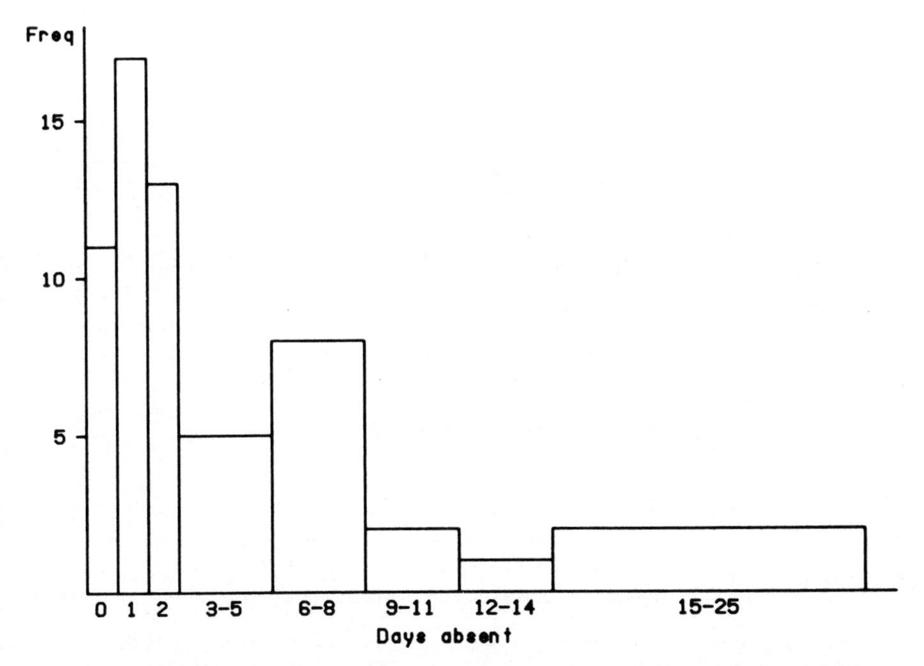

Fig. 2.5 Histogram of number of days absent from school for fictitious sample of children

Notice that the top class is defined here as 15–25 presumably because, in this sample, no absence exceeded 25 days. However, it would be more usual to find this frequency labelled '> 14' or '$\geqslant 15$', i.e. 15 or more days, the upper bound being the total number of school days in the year. For purposes of drawing the histogram, a somewhat arbitrary decision must then be made on the upper limit of the variable concerned; the same situation occasionally arises with lower limits.

The **frequency polygon** is a variant of the histogram constructed by marking where the mid-points of the tops of the histogram boxes would be (if they were to be drawn) and joining these points with straight lines. Thus, for the petrol consumption data (Example 2.4) we would have the picture shown in Fig. 2.6. Frequency polygons are more useful than histograms when we wish to compare two (or more) **frequency distributions** by superimposing one on the other; for example, in Fig. 2.6, the performance of another make of car is shown relative to the first using dotted lines.

The pictorial representation of **cumulative** frequencies is sometimes useful. Such a graph is essentially like a frequency polygon except that cumulative figures (working upwards from low values of the variable) are plotted above the *upper* boundaries of the classes. Such a representation is usually called an **ogive** (originally an architectural term describing similarly shaped vaulted ceilings). Thus, for the school absences of Example 2.5, we have the following cumulative frequency table.

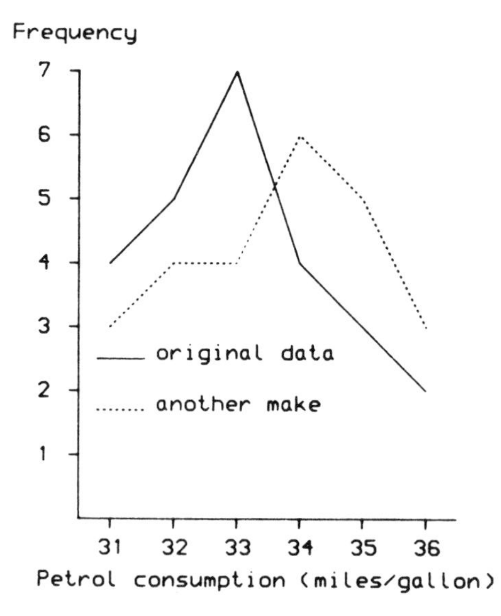

Fig. 2.6 Frequency polygons for petrol consumption of two fictitious makes of car

No. of days absent	Frequency	Cumulative frequency	Cumulative percentage
0	11	11	9.9
1	17	28	25.2
2	13	41	36.9
3– 5	15	56	50.5
6– 8	24	80	72.1
9–11	6	86	77.5
12–14	3	89	80.2
15–25	22	111	100.0

The ogive is shown in Fig. 2.7. It shows what proportion of the pupils in the sample had less than any specified amount of absence.

A type of frequency plot that is, in a way, the mirror image of the ogive is particularly useful where the variable is time or distance. In this case, the left-hand end of the graph corresponds to 100 per cent, this being cumulatively *depleted* by the frequencies in successive classes. For example, the data might relate to times to breakdown of machines, lengths of service of employees or remaining lifetimes

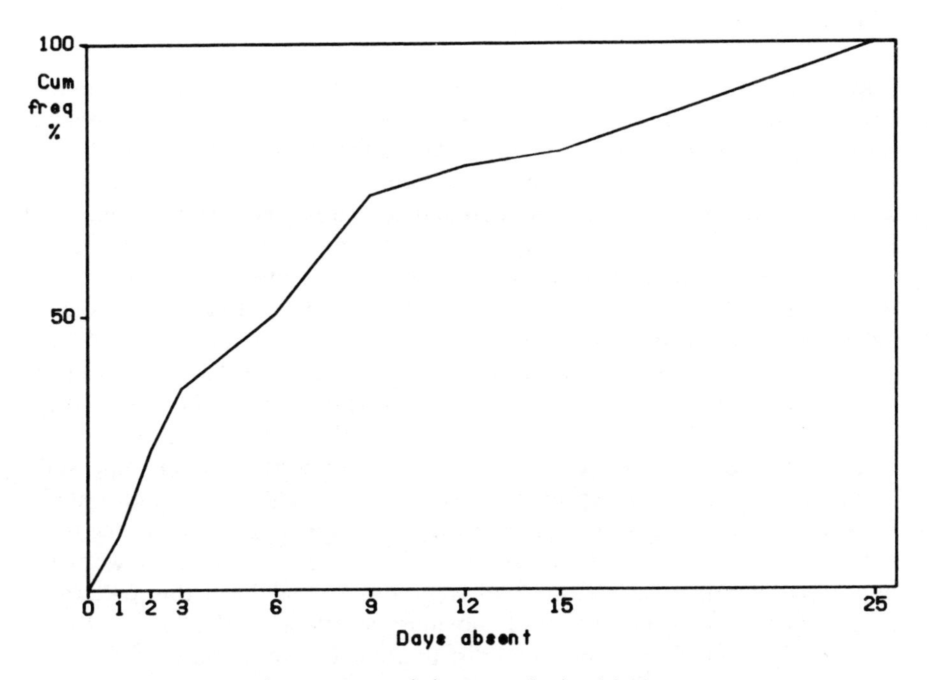

Fig. 2.7 Ogive of absences of schoolchildren

of disease sufferers; not surprisingly, this kind of graph is called a **survivor plot**, the number or proportion still 'alive' at the beginning of a class interval being plotted above the *left-hand* end of that class. We leave the reader to draw the survivor plot for the petrol consumption data (Example 2.4) from the following reformulation:

Miles/gallon	Frequency	Percentage that 'survived' to start of distance class
30.5–31.4	4	100.0
31.5–32.4	5	84.0
32.5–33.4	7	64.0
33.5–34.4	4	36.0
34.5–35.4	3	20.0
35.5–36.4	2	8.0

2.5 POPULATION DISTRIBUTIONS FOR CONTINUOUS VARIATES

(The material in this section is not needed elsewhere in this book but may be helpful in extending the reader's understanding of the pictorial representation of variability.)

The frequency polygon for sample measurements of a continuous random variable can be thought of as a single realization from the infinite number of possible polygons (for different data and/or different choice of classes). We can suppose that, for the population from which the sample is drawn, the distribution of values of the variable could be represented by a polygon with an infinity of infinitesimally narrow classes. Such a polygon would, of course, be a smooth curve with a considerable diversity of possible shapes, of which Fig. 2.8 gives a few examples. The vertical axis still represents the relative frequency of variate values on the horizontal axis.

Figure 2.8(a) shows a so-called **normal** curve, a pattern that is particularly important, partly because most biological variables follow this distribution. However, notice that, in this conceptualization, the variable concerned could (even if only rarely) be negative which in most situations would be impossible.

Figure 2.8(b) gives an **exponential** distribution – the pattern appropriate to intervals between events that are occurring at random. Examples are the times between radioactive particle emissions or lengths of yarn spun between the appearances of flaws. (That such observations should have this pattern of variability is perhaps not intuitively obvious; nevertheless, it is so.)

Figure 2.8(c) exhibits a **lognormal** distribution. The variation in income in a country is often of this form, relatively few people having very high incomes; the pattern applies similarly to many other aspects of economic 'size'. Some

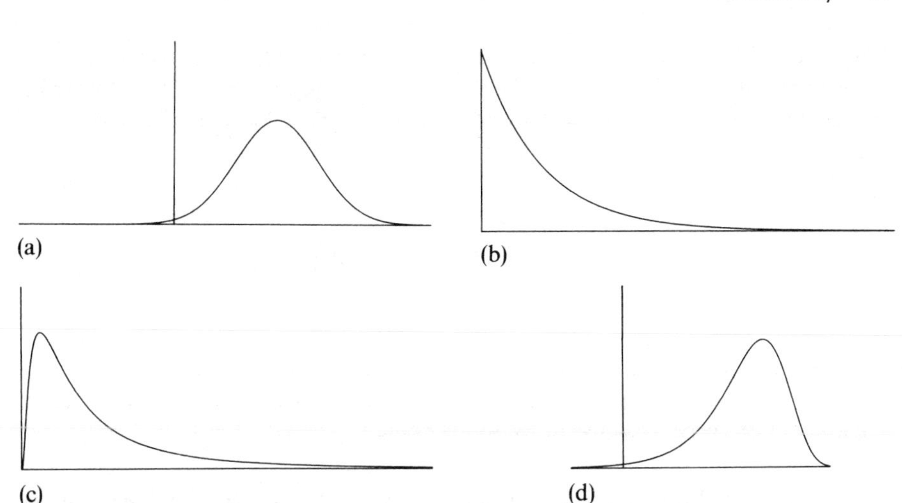

Fig. 2.8 Some population distributions for continuous variables: (a) normal distribution, (b) exponential distribution, (c) lognormal distribution, (d) extreme value distribution

biological measurements such as certain human blood chemicals may also have this form of scatter, the infrequent high values possibly appearing because of ill health. Particle size in the gravel of a river bed is an example from another application area. If we take logarithms of data having a lognormal distribution, the resulting values will have a normal distribution.

Figure 2.8(d) shows an **extreme value** distribution. This kind of pattern has been found useful in the analysis of wind speed, earthquake magnitude, depth of pitting in pipeline corrosion, and so on. Once again, negative values will not usually be admissible; to that extent, the mathematical formulation must be an approximation to reality.

Figure 2.8(a) is an example of a (bell-shaped) **symmetric** distribution. The others represent **skew** distributions, Figs 2.8(b) and 2.8(c) being **positively** skewed (i.e. having a long **tail** to the right) whereas Fig. 2.8(d) is **negatively** skewed. Usually it is the tails of distributions that are of greatest interest since they tell us about the proportion of values that exceed (right-hand tail) or are exceeded by (left-hand tail) any specified figure. Naturally, population analogues of the ogive and survivor plot are relevant to this.

Remember, the above are only examples. For instance, there are a number of bell-shaped symmetric distributions other than the normal and they cannot readily be differentiated by eye.

SUMMARY

Random variables represent qualitative or quantitative attributes of sample or population members. The numbers in qualitative categories can be presented in frequency tables or, graphically, in bar charts or pie diagrams. Similar devices can be used for discrete quantitative data with only a moderate number of possible values. The range of a continuous random variable must be segmented arbitrarily into classes to provide histogram or frequency polygon summaries.

Understanding of nature can often be assisted if the variability in data can be related to a theoretical model as represented by a population frequency distribution.

EXERCISES

(These exercises cover a variety of fields in real situations. For each, decide what type of variate is involved–qualitative, continuous, discrete, etc. Remember that a picture is supposed to be 'worth a thousand words'. Is yours?)

1. In a study of 31 224 patients on surgical waiting lists in Wales and the West Midlands of England (Davidge *et al.*, 1987) the data shown in Tables E2.1 and E2.2 were reported.

Table E2.1 Patients on surgical waiting lists for whom age was recorded

	Age	
Specialty	under 65 yr	65 yr and over
General surgery	5933	1277
Orthopaedics	6483	1676
ENT	7724	155
Gynaecology	2557	91
Ophthalmology	1879	2597

Table E2.2 Patients on surgical waiting lists for whom date placed on list was known

	Waiting period	
Specialty	under 1 yr	1 yr and over
General surgery	4035	3441
Orthopaedics	4658	3685
ENT	3777	4235
Gynaecology	1688	1055
Ophthalmology	2877	1748

(a) Present the findings in an informative way, possibly making use of graphical methods.
(b) Neither table includes all the patients in the study. Discuss the biases that this may introduce.
(c) Compose a few clear but concise sentences that describe the main features of these waiting lists.

(Davidge, M., Harley M., Bickerstaff, L. and Yates, J. (1987) The anatomy of large inpatient waiting lists. *Lancet*, **1987**, i, 794–796.)

2. Lancaster (1972) discusses strike figures for all stoppages lasting at least a day and involving at least ten employees in 1965.
   (a) For a large group of industries, there were 373 strikes over wage claims, 241 were due to disputes over employment of particular persons, 132 were the result of arguments over working arrangements and there were 94 other stoppages. Display this information graphically.
   (b) Strike durations (in days rounded upwards) for the construction and transport industries are shown in Table E2.3.

Table E2.3 Duration and number of strikes in construction and transport industries.

Construction		Transport	
Duration (days)	No. of strikes	Duration (days)	No. of strikes
2	44	2	50
3	33	3	19
4	28	4	10
5	23	5	5
6	11	6	2
7	12	7–20	13
8	9	> 20	3
9	6		
10	13		
11–15	16		
> 15	30		

Compare strike duration in these two industries by means of two histograms. What proportions of stoppages last more than five days in the two cases? Show these on your diagrams by means of shading.

(Lancaster, T. (1972) A stochastic model for the duration of a strike. *J.R. Statist. Soc.*, A, **135**, 257–271.)

3. Burrell and Cane (1982) give borrowing frequencies for Bath University Library and the long-loan section of Sussex University Library, shown in Table E2.4.

Table E2.4 Borrowing from university libraries.

No. of occasions borrowed	No. of books	
	Bath	Sussex
1	17 282	9674
2	4 776	4351
3	1 646	2275
4	496	1250
5	132	663
6	95	355
7	24	154
8	12	72
9	2	37
10	1	14
11	0	6
12	0	2
13	0	0
14	0	1

(a) Compare these two borrowing patterns graphically and comment.
(b) Data are also given from the Sussex short-loan library for two consecutive time periods. The borrowing frequencies during the second time period for those items never borrowed in the first period and those borrowed ten times in the first period

Table E2.5 Borrowing from Sussex University short-loan library in successive time periods

No. of occasions borrowed in second period	Never borrowed in first period	Borrowed ten times in first period
0	849	30
1	445	22
2	290	51
3	198	60
4	127	87
5	85	87
6	38	89
7	27	101
8	21	106
9	8	97
10	7	90
11	8	70
12	5	50

Table E2.5 (*Contd.*)

No. of occasions borrowed in second period	Never borrowed in first period	Borrowed ten times in first period
13	3	44
14	3	48
15	2	35
16	1	18
17	2	14
18	3	11
19	1	5
20	3	12
>20	1	17

were as shown in Table E2.5. Compare these distributions with each other and with the previous two and comment.
(Burrell, Q.L., and Cane, V.R. (1982) The analysis of library data, *J.R. Statist. Soc.*, A, **145**, 439–471.)

4. The serum creatine kinase (SCK) levels in the blood of 120 healthy women studied by Skinner *et al.* (1975) can be summarized as shown in Table E2.6 (in international units per litre):

Table E2.6 SCK levels in blood of healthy women

Level	Numbers of women
5–10	1
10–15	7
15–20	18
20–25	33
25–30	23
30–35	17
35–40	4
40–45	4
45–50	5
50–55	1
55–60	5
60–65	0
65–70	2

(a) Draw the frequency polygon for these data.
(b) The SCK levels for 30 female carriers of Becker muscular dystrophy were:

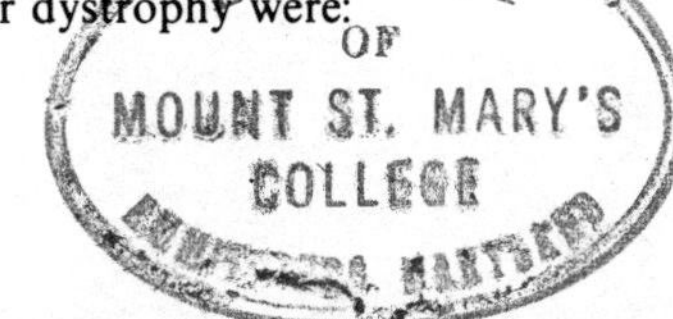

30.0	133.0	131.0	176.0	107.0	114.0	20.0	97.5	51.0	140.5
29.5	46.0	58.0	41.0	79.0	55.5	53.0	30.5	84.5	111.0
10.0	44.0	96.0	53.2	42.0	15.0	50.0	27.0	51.0	34.0

What are your first impressions about these measurements? Superimpose the frequency polygon for these carriers on the previous diagram.
Describe the two polygons and their relationship in words. What proportion of carriers have SCK levels above the highest found among the (presumably) non-carrier women?

(Skinner, R., Emery, A.E.H., Anderson, A.J.B. and Foxall, C. (1975) The detection of carriers of benign (Becker-type) X-linked muscular dystrophy. *J. Medical Genetics*, **12**, 131–134).

5. Ollerenshaw and Flude (1974) studied a random sample of 288 women teachers who had returned to full-time teaching during 1971–72 after at least a year out of full-time employment as teachers.
   (a) The number of years in first teaching post was summarized as shown in Table E2.7.

Table E2.7 Time in first post of teachers subsequently leaving and re-entering full-time teaching

Number of years	Numbers of graduates	Numbers of non-graduates
<1	0	2
1–2	20	6
2–3	28	10
3–4	30	6
4–5	26	7
5–6	27	5
6–8	45	9
8–10	31	4
10–15	23	3
>15	5	1

Draw superimposed frequency polygons of the first employment experiences of these teachers, explaining how you have dealt with the longest-serving group.
   (b) Why can these figures not provide an estimate of the proportion of teachers who left within, say, three years of starting their first-jobs? (*Hint*: Who are missing from the study?)
   (c) For those who had taught part-time before returning to full-time, the part-time period is summarized in Table E2.8.

Table E2.8 Years spent in part-time teaching

Number of part-time years	Numbers of teachers
<1	78
1–2	39
2–3	17
3–4	8
4–5	9
5–6	8
6–7	4
7–8	2
>8	5

Draw a frequency polygon for years of part-time employment.

(d) The ages at time of returning to full-time teaching were distributed as in Table E2.9.

Table E2.9 Age at return to full-time teaching

Age (years)	Numbers	Age (years)	Numbers
23	1	38	14
24	2	39	16
25	5	40	15
26	3	41	11
27	10	42	9
28	10	43	7
29	12	44	10
30	13	45	5
31	15	46	5
32	15	47	2
33	23	48	3
34	12	49	2
35	23	50	1
36	18	>50	11
37	15		

Draw a frequency polygon for the age distribution.

(e) Describe in words the various graphs you have drawn.

(Ollerenshaw, K. and Flude, C. (1974) *Returning to Teaching*, University of Lancaster.)

6. A botanist measured the lengths, breadths and length:breadth ratios for the leaves of 32 herbarium specimens of *Rhododendron cinnabarinum*, a species that extends the length of the Himalayan range. These data are given in Table E2.10

Table E2.10 Measurements of leaves from *Rhododendron cinnabarinum*

Length (l)	Breadth (b)	l/b	Length	Breadth	l/b
69	19	3.63	43	28	1.54
26	25	1.04	38	37	1.03
39	30	1.30	28	24	1.17
68	25	2.72	34	27	1.26
21	34	0.62	87	30	2.90
59	18	3.28	83	26	3.19
36	34	1.06	47	37	1.27
72	23	3.13	44	26	1.69
59	20	2.95	33	37	0.89
58	36	1.61	69	26	2.65
40	30	1.33	32	28	1.14
39	32	1.22	60	19	3.16
72	22	3.27	48	34	1.41
36	21	1.71	51	32	1.59
47	35	1.34	42	35	1.20
19	24	0.79	49	25	1.96

(a) Construct histograms for the lengths, breadths and ratios. Do any of these support a suggestion that the species may be divisible into two subspecies? Explain your answer in terms of leaf shape.

(b) Even if you have no botanical expertise,
  (i) Suggest other possible explanations for the pattern.
  (ii) Do you think the specimens represent a random sample of *Rhododendron cinnabarinum* plants?

(See Cullen, J. (1977) Work in progress at Edinburgh on the classification of the genus *Rhododendron. Year Book of the RHS Rhododendron and Camellia Group*, pp. 33–43.)

7. The volumes of 500 hand-made wooden beer casks were measured by weighing each empty and then full, the machine having a dial calibrated in quarter-pints above the required minimum capacity. The casks were independently measured by a second operator employed to extract those less than two, or six or more pints above the minimum, lifting them onto a lorry for return to the makers. The measurements are given in Table E2.11

Table E2.11 Estimation of beer cask volumes

Pints above min. req.	All casks	Accepted casks	Pints above min. req.	All casks	Accepted casks
0.00	2	1	5.00	52	51
0.25	0	0	5.25	16	16
0.50	3	1	5.50	39	38
0.75	0	0	5.75	6	6

Table E2.11 (*Contd.*)

Pints above min. req.	All casks	Accepted casks	Pints above min. req.	All casks	Accepted casks
1.00	8	1	6.00	53	5
1.25	2	0	6.25	9	3
1.50	8	4	6.50	11	1
1.75	1	0	6.75	8	1
2.00	16	14	7.00	30	1
2.25	2	2	7.25	4	1
2.50	17	16	7.50	13	2
2.75	9	8	7.75	3	0
3.00	40	38	8.00	17	0
3.25	6	6	8.25	1	1
3.50	25	23	8.50	3	0
3.75	4	4	8.75	3	0
4.00	35	34	9.00	2	0
4.25	8	8	9.25	0	0
4.50	32	31	9.50	2	1
4.75	9	9	9.75	1	0

(a) Draw a histogram of the measurements for all casks as recorded, i.e. using quarter-pint classes. Comment.
(b) Re-draw the histogram using one-pint classes.
(c) Superimpose on the second diagram the histogram of the casks classed as acceptable. Comment.
(d) What proportions of casks in this sample were rejected as being
   (i) too high
   (ii) too low?

Suggest a sensible adjustment to the rejection rule that would bring these roughly into balance and give an overall percentage acceptable of about 80%? What might be the financial implications of this change?

(See Cunliffe, S.V. (1976) Interaction. *J. R. Statist. Soc.*, A, **139**, 1–16, for a discussion of similar beer cask data.)

8. Daldry (1976) reported on yearly maxima of wind gusts (miles per hour) at Edinburgh and Paisley as follows:

Edinburgh (1918–67)

78	81	66	79	76	74	75	85	73	81
81	86	87	76	66	69	65	85	74	87
66	74	82	73	78	69	75	68	76	68
76	81	75	75	79	74	85	87	69	67
88	73	71	68	78	83	79	88	71	73

Paisley (1914–67)

66	64	63	59	64	63	69	63	59	63
67	67	63	90	68	55	68	64	64	63
67	73	83	49	70	59	73	62	58	63
62	55	66	59	69	69	55	72	63	55
68	66	67	75	49	63	60	70	77	56
62	62	59	59						

(a) Display this information graphically (5 mph classes, perhaps) and compare the two patterns in a few words.

(b) Draw the ogives, one superimposed on the other. Add the horizontal line that would cut off the top 25% of wind gusts. Roughly what wind speeds would correspond to this at the two sites?

(Daldry, S. J. (1976) Some extreme value problems in the analysis of meteorological data. PhD Thesis, University of Hull.)

9. Gore (1981a) reports on the 3922 breast cancer patients first diagnosed at the Western General Hospital in Edinburgh between 1954 and 1964.

(a) Tumour sizes at diagnosis were extracted from case notes and grouped as shown in Table E2.12.

Table E2.12 Breast tumour size

Tumour size (cm)	Number of patients
too small to measure	24
0–1	148
1–2	455
2–3	835
3–4	727
4–5	519
5–6	308
6–7	189
7–8	162
8–9	80
9–10	89
> 10	163
not recorded	223

Draw the histogram of tumour size for these patients. Shade that part of the histogram representing tumours larger than 5 cm. What proportion of all tumours do these represent?

Comment on the effect of the 'size unrecorded' cases.

(b) The 347 patients diagnosed during 1956 were followed up for 20 years with the pattern of deaths shown in Table E2.13 (Gore 1981b).

Table E2.13 Survival of breast cancer patients

Survival (years)	Number dying	Survival (years)	Number dying
< 1	62	10–11	8
1–2	45	11–12	8
2–3	38	12–13	9
3–4	28	13–14	5
4–5	25	14–15	2
5–6	10	15–16	3
6–7	14	16–17	4
7–8	11	17–18	7
8–9	9	18–19	1
9–10	8	19–20	3

(Notice that 47 of the original patients were still alive at the end of the 20-year study. The observations on these patients are said to be (right) **censored**.)
Construct the frequency polygon for the survival times of the 347 patients. What proportion of patients survived for at least ten years after diagnosis?
Draw the survivor plot for the data. Add the horizontal line that cuts off survival times of ten years or longer.

(Gore, S.M. (1981a) Analysis of survival data in breast cancer. PhD Thesis, University of Aberdeen.
Gore, S.M. (1981b) Assessing methods – descriptive statistics and graphs. *Br. Med. J.*, **283**, 486–488.)

10. In a comparison of the literary styles of G.B. Shaw and G.K. Chesterton, Williams (1939) counted the numbers of words in 600 sentences sampled from a major work of each. The frequencies were as shown in Table E2.14.
    (a) For GBS, the first 15 sentences in each of the first 40 sections of the work were chosen (though the frequencies add to only 598). For GKC, the first 30 sentences were counted in each of the first 20 chapters. Are you happy with such a sampling scheme? Suggest a better method.
    (b) Draw histograms for the two distributions of sentence length, giving particular thought to sensible class widths. Comment.
    (c) It is sometimes argued that sentence length has a lognormal distribution (see Fig. 2.8(c)). Why is this, in a strict sense, impossible? If we were to accept the claim with reservations, then the logarithm of sentence length would have a normal distribution (Fig. 2.8(a)). To check this by drawing the resulting histograms, you could simply take logarithms of your present class boundaries so that the class frequencies would be the same; unfortunately the histograms would then be more difficult to draw because of the unequal class widths. Alternatively (supposing you are using natural logarithms), you could have class boundaries such as ... 3.2, 3.3, 3.4, 3.5, ... and aggregate the frequencies for corresponding sentence lengths; but the three classes just exemplified would correspond to 2, 4 and 3 sentence lengths respectively. Look at what happens at this point to the resulting class frequencies for both authors. Ponder on this.

Table E2.14 Numbers of words in sentences from works by G.B. Shaw (GBS) and G.K. Chesterton (GKC)

	Frequency			Frequency			Frequency	
Length	GBS	GKC	Length	GBS	GKC	Length	GBS	GKC
1	0	0	24	13	22	47	4	3
2	0	0	25	13	19	48	5	2
3	2	0	26	10	28	49	3	2
4	3	0	27	13	21	50	4	8
5	14	3	28	14	22	51	9	3
6	10	3	29	13	21	52	7	2
7	10	5	30	14	17	53	10	2
8	11	2	31	16	21	54	8	0
9	16	11	32	9	19	55	2	2
10	15	6	33	9	11	56	1	2
11	9	7	34	11	8	57	5	0
12	6	17	35	12	6	58	4	2
13	14	16	36	5	5	59	3	1
14	16	16	37	6	13	60–69	16	1
15	21	15	38	5	6	70–79	15	1
16	12	16	39	9	7	80–89	13	0
17	15	20	40	5	10	90–99	6	1
18	22	28	41	5	6	100–110	2	0
19	12	24	42	6	6	110–119	1	0
20	16	25	43	7	8	120–129	0	0
21	12	21	44	5	4	130–139	0	0
22	17	26	45	5	4	140–149	2	0
23	11	19	46	8	4			

(Williams, C.B. (1939) A note on the statistical analysis of sentence-length as a criterion of literary style. *Biometrika*, **31**, 356–361.)

3

Numerical summary of data – sample statistics

3.1 MEASURES OF LOCATION

A frequency distribution, expressed in tabular or graphical form, provides some degree of reduction of the data. We can go somewhat further by calculating what are termed **sample statistics**. A sample statistic is any quantity derivable from the observed values of a random variable, e.g. the proportion of smokers in a sample of students is itself a random variable since its value will change from sample to sample.

3.1.1 Percentiles

Suppose that in a class of $n = 150$ students, you have a mark of 52%. Is this good? One way of indicating the status of such a mark would be to give the ordinal measure of your position in the class, e.g. 45th from bottom. This would mean that $45/150 = 3/10$ of the class had a mark equal to or less than 52. We would call 52 the 30th **percentile**, i.e. the mark at or below which 30% of the class lie. A few such percentiles will provide a useful description of the frequency distribution – often the **deciles**, 10%, 20%, ..., 90% 100%, or even just the **quartiles**, 25%, 50%, 75%, are sufficient. These are all sample statistics. The most important of all is the 50% percentile, more usually called the **median.**

3.1.2 Median

The median is the value exceeded by 50% of the observations – it is the half-way point of the observed values and obviously tells us something about the *location* of the frequency distribution, i.e. that the marks are distributed round about 73, say. If the number of observations, n, is odd, then the median is the value of the middle observation i.e. the $((n + 1)/2)$th in magnitude. The reader can check that, in Example 2.4, the median petrol consumption is 33.1 miles. When n is even, the median is taken to be the average of the *two* middle observations. Thus, for the part-time teachers of Example 2.3, the median number of children is the average of the tenth and eleventh smallest number of children; since both these are 2, the median is 2.0 children.

3.1.3 Mode

For data grouped into classes, we can also provide a rough indication of the location of a frequency distribution by finding the 'most popular class', i.e. the point below the highest peak of the frequency polygon. We call this point the **mode**. Thus, for the petrol data of Example 2.4, the mode is 33.0 miles, i.e. the mid-point of the 32.5 to 33.4 class. Of course a frequency distribution can have two (or more) almost equally popular classes – it is then said to be **bimodal** (or multi-modal). However, the data on school absences (Example 2.5) would be regarded as unimodal with the mode at 1.5 days.

3.1.4 Arithmetic mean or average

This measure of location is so well known it hardly needs description – it is just

$$\text{mean} = \frac{\sum \text{observations}}{n}$$

where $\sum$ is a symbol meaning 'the sum of'. Thus, for the data of Example 2.4, the mean petrol consumption is

$$\frac{33.2 + 35.1 + 31.4 + \cdots + 34.1 + 34.0}{25} = \frac{827.7}{25}$$
$$= 33.11$$

(Means are usually quoted to one more place of decimals than the raw data.) We could also write this as

$$\frac{30.6 \times 1 + 30.9 \times 1 + 31.2 \times 1 + 31.4 \times 1 + 31.6 \times 1 + 31.8 \times 3 + \cdots}{25}$$

i.e. $30.6 \times \frac{1}{25} + 30.9 \times \frac{1}{25} + 31.2 \times \frac{1}{25} + 31.4 \times \frac{1}{25} + 31.6 \times \frac{1}{25} + 31.8 \times \frac{3}{25} + \cdots$

i.e. $\sum$(observed value × relative frequency of that value)

This is not very useful for calculation but it is a helpful way to think about averages and we can now see how to get a mean from frequency data – we take (see Section 3.5.2)

$$\sum(\text{class mid-point} \times \text{relative frequency in class})$$

Corresponding to each observation in a sample, there will be a **deviation** from the sample mean. It is an interesting property of the mean that the sum of these deviations is zero, since

$$\begin{aligned}\sum \text{deviations} &= \sum (\text{obsn} - \text{mean})\\ &= \sum \text{obsns} - n \times \text{mean}\\ &= \sum \text{obsns} - \sum \text{obsns} = 0\end{aligned}$$

Example 3.1
Suppose we have 5 observations:

$$7,8,3,7,5$$

$$\text{Then mean} = \frac{7+8+3+7+5}{5} = 6.0$$

and the deviations are 1, 2, − 3, 1, − 1, the sum of which is zero. Another interesting property is that the sum of squared deviations from the mean is smaller than the sum of squared deviations from any other value.

	Squared deviations from–				
Obsn	4	5	6	7	8
7	9	4	1	0	1
8	16	9	4	1	0
3	1	4	9	16	25
7	9	4	1	0	1
5	1	0	1	4	9
$\sum$	36	21	16	21	36

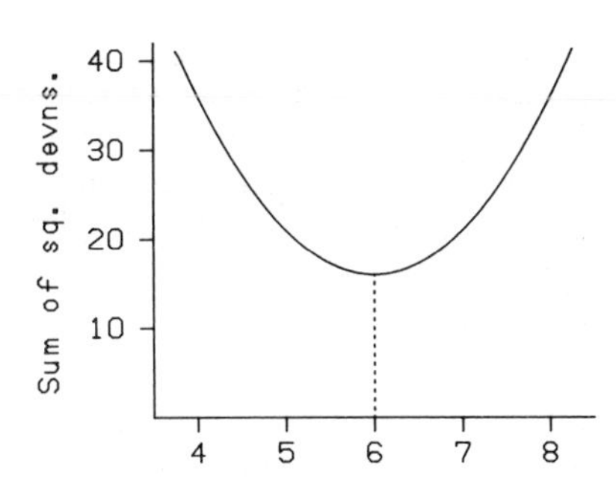

3.1.5 Weighted mean

An important extension to the simple arithmetic average introduced in the previous section is the **weighted mean**. With each observation there is associated some 'weight', and the weighted mean is defined as

$$\frac{\sum(\text{obsn} \times \text{weight})}{\sum \text{weight}}$$

Thus the ordinary mean is a weighted mean with all weights equal to 1; means calculated from grouped data (to be discussed in Section 3.5.2) have weights equal to the class frequencies. However, weighted means find fairly wide application, as will be seen in Chapters 8, 10 and 12.

3.2 WHICH MEASURE OF LOCATION?

In the case of a parent distribution with perfect symmetry, the mode, median and mean for the population coincide. When the distribution is skew either positively or negatively, the mean lies away from the median in the direction of the long 'tail' of the distribution (see Fig. 2.8). In such cases, the sample median may be a more appropriate measure of the location of the distribution than the sample mean. However, for distributions that are not too skew, the sample mean is preferred because

1. it is easy to calculate,
2. properties such as those mentioned in Section 3.1.4 are found to be useful,
3. if repeated samples were taken from a population, we would find that the mean varied much less than the median,
4. in sufficiently large samples, its pattern of variability is always that of the normal distribution (Fig. 2.8(a)) irrespective of the distribution of the observations.

The sample mode is little used because it depends so much on arbitrary class boundaries.

3.3 VARIANCE AND STANDARD DEVIATION

We now have some ideas about how to measure the location of a distribution. We also need to measure the degree of spread of the observations. We could use the **range** of the observed values but that is rather sensitive to the appearance of odd high or low values in the data. Instead, we make use of the deviations from the sample mean. The average of these deviations is no use since they sum to zero. We could use the average of the deviations each taken with a positive sign; this is called the **mean deviation** but is seldom applied, having less satisfactory properties than a measure of spread called the **variance**.

To calculate the variance, the deviations are squared before being added up. This means that there are no problems about sign and, furthermore, large deviations are given extra importance, e.g. a deviation of 1 contributes 1 to the sum, a deviation of 2 contributes 4. This sum of squared deviations (often called the **corrected sum of squares**) is then divided by the number of observations less 1, i.e. $(n-1)$, to form an average. We can explain the use of $(n-1)$ rather than n by observing that, once $(n-1)$ of the deviations are given, the nth is known since they must sum to zero – we say there are $(n-1)$ **degrees of freedom**. So, although all the squared deviations go to make up the sum of squares, we are really looking at the mean of only $(n-1)$ independent items. Notice that the variance is in squared units, e.g. if we are measuring height in centimeteres, the variance of a sample of heights will be in centimetres squared.

The **standard deviation** is just the square root of the variance and is therefore in the same units as the data. Thus, for the five observations of Example 3.1 we have

$$\text{sum of squared deviations} = 16$$
$$\text{hence variance} = \tfrac{16}{4} = 4.0$$
$$\text{and standard deviation} = \sqrt{4} = 2.0$$

It is the standard deviation that is normally quoted as the measure of spread of a distribution. For large samples, it is frequently found that the range is about four times the standard deviation.

3.3.1 Calculating corrected sum of squares

A sum of squared deviations is hardly ever found by actually calculating the deviations, squaring them and adding up – it is too laborious. A formula that can be shown to be equivalent and which is much easier to use is

$$\sum \text{obsns}^2 - \text{CFM}$$

where CFM (the so-called correction for the mean) can be expressed as

$$\left.\begin{array}{l} \text{total}^2/n = 30 \times 30/5 = 180 \\ \text{or} \\ \text{mean} \times \text{total} = 6 \times 30 = 180 \\ \text{or} \\ n \times \text{mean}^2 = 5 \times 6 \times 6 = 180 \end{array}\right\} = \text{CFM}$$

for the data in Example 3.1.
The sum of the squared observations is

$$49 + 64 + 9 + 49 + 25 = 196$$

so that

$$\sum \text{obsns}^2 - \text{CFM} = 196 - 180 = 16, \text{ as before.}$$

We look a little more fully at this calculation in Chapter 13.

3.3.2 Interquartile range

For skew distributions where the median is preferred to the arithmetic mean as the measure of location, it would seem anomalous to measure dispersion in terms of deviations from the mean. The statistic that is usually used in such circumstances is the **interquartile range** – simply the difference between the 75th and 25th percentiles (see Section 3.5.1 for an example).

3.4 NOTATION

The observations of a continuous quantitative random variable are often denoted by x and their mean by $\bar{x}$ (pronounced 'x bar')

$$\text{i.e.} \quad \bar{x} = \frac{\sum x}{n}$$

The corrected sum of squares is then $\sum(x - \bar{x})^2$, conveniently represented as S_{xx} and calculated as

$$\begin{aligned} S_{xx} &= \sum x^2 - (\sum x)^2/n \\ &= \sum x^2 - \bar{x} \times (\sum x) \qquad \text{as described in the Section 3.3.1} \\ &= \sum x^2 - n \times \bar{x}^2 \end{aligned}$$

The variance is denoted by s^2, so that

$$s^2 = \frac{\sum (x - \bar{x})^2}{n - 1} = \frac{S_{xx}}{n - 1}$$

and the standard deviation

$$s = \sqrt{\frac{\sum (x - \bar{x})^2}{n - 1}} = \sqrt{\frac{S_{xx}}{n - 1}}$$

3.5 GROUPED DATA

How can we calculate medians, means and variances when only grouped frequencies rather than the raw data are available?

Example 3.2
The marks of 150 students grouped into 5-mark classes are shown in Table 3.1.

Table 3.1 Student exam marks

Marks	Frequency	Cumulative frequency	Cumulative % frequency
35–39	3	3	2
40–44	6	9	6
45–49	12	21	14
50–54	15	36	24
55–59	27	63	42
60–64	30	93	62
65–69	18	111	74
70–74	15	126	84
75–79	12	138	92
80–84	12	150	100

3.5.1 Median

The cumulative frequencies and cumulative percentage frequencies are easily calculated and give the ogive shown in Fig. 3.1.

We can see that the median must lie in the class 60–64, but we do not know exactly where. We can only assume that the marks in this class are evenly spread along the 59.5 to 64.5 interval and therefore, since the 50% point is

$$\frac{50\text{–}42}{62\text{–}42} = \frac{2}{5}$$

of the distance up from the lower boundary of the class on the vertical axis, the

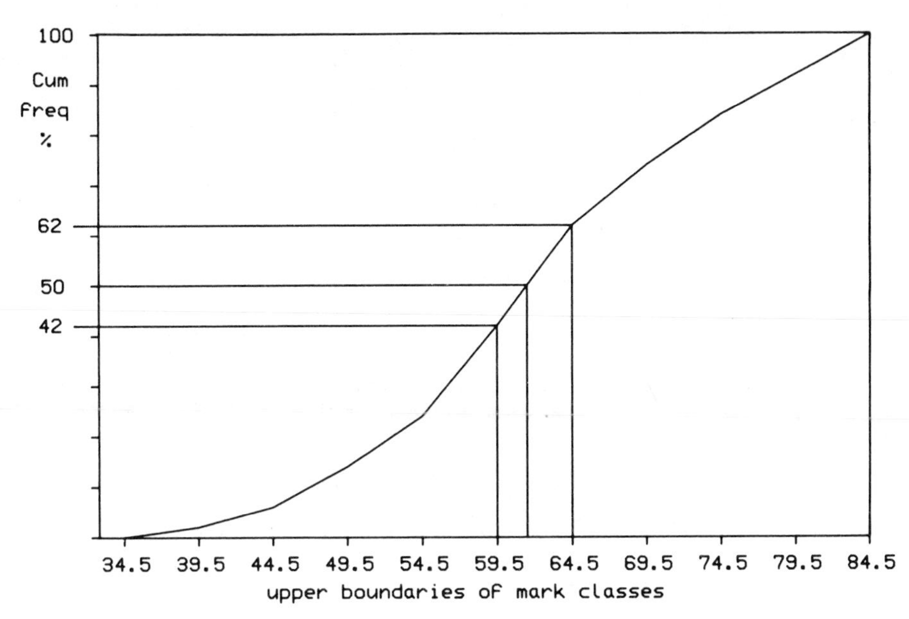

Fig. 3.1 Ogive of fictitious examination marks for 150 students (Example 3.2)

median is estimated to be 2/5 of the way along the 59.5 to 64.5 interval on the horizontal axis

$$\text{i.e.} \quad 59.5 + \frac{2}{5} \times 5 = 61.5$$

(The fancy name for this process is **linear interpolation**.) Thus, in general,

$$\text{median} = \begin{matrix}\text{lower boundary of} \\ \text{median class}\end{matrix} + \left(\frac{50 - \text{low bdry. cum.\% freq.}}{\text{\% freq. in class}} \right) \times \begin{matrix}\text{class} \\ \text{width}\end{matrix}$$

Obviously the same procedure can be used for any percentile, e.g.

$$25\text{th percentile} = 49.5 + \frac{(25 - 24)}{(42 - 24)} \times 5 = 49.8$$

and

$$75\text{th percentile} = 64.5 + \frac{(75 - 74)}{(84 - 74)} \times 5 = 65.0$$

so that interquartile range $= 65.0 - 49.8 = 15.2$

3.5.2 Mean and variance

Contrary to what we assumed for the estimation of the median, we now assume

Table 3.2 Values for calculating mean and variance of exam marks

Class mid-point (v)	Frequency (f)	$f \times v$	$f \times v^2$
37	3	111	4 107
42	6	252	10 584
47	12	564	26 508
52	15	780	40 560
57	27	1539	87 723
62	30	1860	115 320
67	18	1206	80 802
72	15	1080	77 760
77	12	924	71 148
82	12	984	80 688
	$n = 150$	$\sum = 9300$	$\sum = 595\,200$

that all observations in a class are concentrated at the class mid-point, v. Thus, we have values as shown in Table 3.2. Using the appropriate versions of the formulae in Sections 3.1.4 and 3.3.1,

$$\text{mean} = \frac{\sum(f \times v)}{n} = \frac{9300}{150} = 62.0$$

$$\begin{aligned} \text{corrected sum of squares} &= \textstyle\sum(f \times v^2) - (\sum(f \times v))^2/n \\ &= 595\,200 - 9300^2/150 = 18\,600 \\ \text{variance} &= 18\,600/149 = 124.83 \\ \text{standard deviation} &= 11.2 \end{aligned}$$

All of these estimates will almost certainly differ from the figures we would obtain if we had the raw data rather than just a frequency table. It would certainly be meaningless to quote them to further places of decimals.

Notice that where the highest or lowest class in a frequency table is open-ended it may be difficult to compute the mean because of uncertainty on what mid-point to use for such a class. No problem generally attaches to calculation of the median in this situation. (Even the interquartile range can be found unless more than one quarter of the observations lie in the open-ended class.) Luckily, the difficulty is more frequently encountered with skew distributions (e.g. for patient survival time – see Exercise 9(b) of Chapter 2) where the median is, in any case, a better location measure than the mean.

3.6 VARIABILITY OF SAMPLE MEANS

As has already been stated, the sample mean is a random variable and hence has a distribution of values that could appear if repeated samples were to be taken. It follows that the sample mean has an associated standard deviation measuring *its* variability; this statistic is called the **standard error** of the mean.

The standard error of the mean can be computed from the basic variance, s^2, of the variable concerned even though only a single sample has been taken. It can be shown that

$$\text{standard error of } \bar{x} = \sqrt{\left(\frac{s^2}{n}\right)}$$

Thus, for the petrol consumption data of Example 2.4,

$$\sum x^2 = 27\,459.83$$

so that

$$s^2 = \left(27\,459.83 - \frac{827.7}{25}\right)^2 \Big/ 24 = 2.347$$

and hence

$$\text{standard error of mean} = \sqrt{\frac{2.347}{25}} = 0.306\,427$$

For raw data, a standard error (SE) is usually quoted to one more place of decimals than the mean itself, i.e. we could summarize the average petrol consumption as

$$33.11, \quad \text{SE} = 0.306$$

(The notation 33.11 ± 0.306 is sometimes used but this is capable of other interpretations and is best avoided; it certainly does not mean that all observed values lie between 32.804 and 33.416!)

We can now proceed one stage further to consider the comparison of two means. Suppose we measure some quantitative variable, x, for two *independently selected* samples of sizes n_1 and n_2 giving means $\bar{x}_1$ and $\bar{x}_2$ and variances s_1^2 and s_2^2 respectively. Then the difference $(\bar{x}_1 - \bar{x}_2)$ is a sample statistic that would vary over different pairs of samples. The standard error of this difference is

$$\text{SE}(\bar{x}_1 - \bar{x}_2) = \sqrt{\left(\frac{s_1^2}{n_1} + \frac{s_2^2}{n_2}\right)}$$

unless it can be assumed that the two populations being compared are equally variable, in which case

$$\text{SE}(\bar{x}_1 - \bar{x}_2) = \sqrt{s_P^2\left(\frac{1}{n_1} + \frac{1}{n_2}\right)}$$

where

$$s_p^2 = \frac{(n_1 - 1)s_1^2 + (n_2 - 1)s_2^2}{(n_1 - 1) + (n_2 - 1)}$$

(Notice that s_p^2 – called the **pooled** sample variance – is the weighted mean of the two separate sample variances with the degrees of freedom as weights.)

Thus, suppose the petrol consumption of 20 examples of a second make of car were evaluated (by the same driver using the same fuel) to give $\bar{x}_2 = 33.78$ and $s_2^2 = 2.737$. (Remember, $\bar{x}_1 = 33.11$ and $s_1^2 = 2.347$.) The standard error of the difference in means is therefore

$$\sqrt{\left(\frac{2.347}{25} + \frac{2.737}{20}\right)} = 0.480$$

The observed difference in means of 0.67 is not spectacular in the light of this degree of variability; i.e. if the experiment were repeated, the results might be somewhat different. Considerations of this kind are looked at again in Chapter 5. (If we had good reason to believe that the two makes of car were equally variable in performance – the sample variances are at least consistent with such a claim – then we could compute

$$s_p^2 = \frac{24 \times 2.347 + 19 \times 2.737}{24 + 19}$$

and

$$\text{SE}(\bar{x}_1 - \bar{x}_2) = \sqrt{(2.519 \times (\tfrac{1}{25} + \tfrac{1}{20}))} = 0.476)$$

3.6.1 Paired comparison

It must be emphasized that the above standard errors are not appropriate to the case where we wish to compare the means of two variates measured on either the *same* sample units (e.g. body temperature before and after administration of a fever-reducing drug) or units that are associated in pairs (e.g. IQs of husband and wife). In such a situation (called a **paired comparison** study), the correct analysis is to derive a new variable from the difference between the two under consideration; naturally, the mean of this new variable equals the difference between the means of the two original variables and *its* standard error is the correct one to use.

For instance, we could make a paired comparison of two different brands of petrol by running each of the original 25 cars on a gallon of the second fuel. The 25 within-car differences in distance travelled (some probably with a negative sign) provide the appropriate standard error for the difference in average performance of the two brands.

SUMMARY

Sample statistics provide quantitative summary measures of the information in data. Proportions in different categories are appropriate for qualitative variables.

For symmetrically distributed quantitative data, the mean is the best indicator of the middle of the spread of observed values while the standard deviation measures the extent of the dispersion of observations about that average. Both can be computed for grouped data. The weighted mean is an important extension of the simple average. For quantitative observations from a skew distribution, the median and interquartile range are better measures of location and dispersion respectively; again, both can be computed for grouped data. The variability of an arithmetic mean or the difference of two means is measured by the appropriate standard error.

EXERCISES

1. For any of the sets of data given in Exercises 2–10 of Chapter 2, calculate appropriate measures of location and spread. Check that your answers make sense and that you understand what they are telling you. (If you have time, you may care to compare the computation of corrected sums of squares by summing the squared deviations from the mean with the more practical method given in Sections 3.3.1 and 3.5.2.)

2. Table E3.1 gives the age distribution of the populations of the United Kingdom and Scotland in 1971.

Table E3.1 Age distributions for UK and Scotland, 1971

	Numbers in thousands	
Age	UK	Scotland
Under 5	4 505	44.3
5–14	8 883	910.5
15–29	11 679	1099.1
30–44	9 759	912.7
45–64	13 383	1217.9
65–74	4 712	426.5
75 and over	2 594	218.0

Source: 1971 Census.

(a) Compare the mean and median ages for both Scotland and for the United Kingdom. Comment on the validity of any assumptions you make.
(b) Estimate the proportions of the United Kingdom and Scottish populations that were aged 16 and over. Are your answers likely to be too high or too low?

3. (a) Humans are genetically constrained to metabolize sulpha-drugs at one of two distinct rates. Confirm this by considering data reported for 58 healthy subjects by Price Evans and White (1964) giving the percentage of sulphamethazine excreted in the urine during the 8 hours after ingestion, which is shown in Table E3.2. Draw the histogram for these data and find the means and their standard errors for both 'slow' and 'fast' metabolizers.

Table E3.2 Excretion of sulphamethazine in urine

% excreted	Number	% excreted	Number
40–42	2	66–68	1
42–44	2	68–70	2
44–46	6	70–72	3
46–48	6	72–74	4
48–50	4	74–76	1
50–52	4	76–78	3
52–54	5	78–80	3
54–56	0	80–82	4
56–58	0	82–84	2
58–60	0	84–86	0
60–62	0	86–88	1
62–64	1	88–90	2
64–66	2		

(b) In the same study, liver samples from 8 'slow' and 5 'fast' metabolizers were removed during abdominal operations. These samples were separately immersed in standard solutions of sulphamethazine and the amounts (in μg) of the drug remaining almost immediately after immersion and two hours later were assayed to be:

'Slow'

0 hours	48.1	48.3	48.1	48.5	50.9	51.0	48.7	48.8
2 hours	45.2	47.7	49.1	44.0	48.9	48.5	43.8	44.2

'Fast'

0 hours	20.0	41.8	28.7	34.0	35.5
2 hours	17.5	27.9	7.0	18.7	18.6

Find the means and their standard errors for the four sets of measurements. (By the way, the standard error for the 'slow at 2 hours' mean appears to be incorrectly reported in the original paper.)

(c) For each group of subjects, compute the standard error for the difference in mean sulphamethazine levels at the two times.

Compute the standard error appropriate to the comparison of the time difference means in the two groups of subjects.

Comment on the experiment.

(Price Evans, D.A. and White, T.A. (1964) Human acetylation polymorphism. *J. Lab. and Clin. Med.*, **63**, 394–403.)

4. Roughly 50% of common crossbills have bills with upper mandibles crossing to the left (left-billed), the remainder having upper mandibles crossing to the right (right-billed). To investigate whether the method of feeding places more strain on one leg than the other (dependent on the direction of crossing), Knox (1983) measured the lengths (in

mm) of the left and right tarsal bones of 12 specimens of each kind of bird. The data are shown in Table E3.3.

Table E3.3 Lengths of tarsal bones in crossbills

Common crossbill (*Loxia curvirostra curvirostra*)					
Left-billed			Right-billed		
Bird	Left leg	Right leg	Bird	Left leg	Right leg
1	19.0	17.4	13	17.8	18.0
2	17.1	17.3	14	15.6	17.1
3	16.6	16.6	15	16.9	17.2
4	17.0	16.7	16	16.9	17.1
5	18.3	18.2	17	17.0	17.3
6	17.5	17.3	18	16.9	17.1
7	16.4	16.3	19	17.4	17.6
8	17.1	16.9	20	17.4	17.7
9	17.5	16.7	21	17.7	17.7
10	17.0	16.8	22	16.5	17.5
11	18.6	17.5	23	17.1	17.3
12	16.3	16.2	24	16.7	16.9

(a) Find the four means (left and right tarsals for each type of bird) and their associated standard errors.
(b) For each type of bird find the mean difference in tarsal length and associated standard error.
(c) Compare the right-minus-left tarsal mean difference in left-billed birds with the left-minus-right tarsal mean difference in right-billed birds. What is the standard error for this comparison?

(Knox, A.G. (1983) Handedness in crossbills (*Loxia* and the Akepa *Loxops coccinea*. *Bull. Br. Ornithologists' Club*, **103**, 114–118.)

5. 'In Scotland 44% of adults over the age of 16 have no natural teeth. Our teeth are rotting at the rate of one every $2\frac{1}{2}$ seconds'. (Health Education poster of about 1976.)
'Dentists in the UK remove 10 tons of teeth a year because teeth are not properly cared for'. (*Sunday Times*, 3/10/76.)
Use this information and the table in Exercise 2 to estimate the average weight of a tooth in grams. Make a full and critical list of all the assumptions you are making. If you feel your answer is unreasonable, suggest what might be wrong.

6. (A slightly mathematical brain teaser.)
The mean and standard deviation of a set of eight observations are reported as 12.00 and 1.99 respectively. Six of the original recordings are known to have been

10.3 12.4 9.2 14.1 14.6 11.2

but the remaining two have been lost. Find them.

4
Association between variables

4.1 INTRODUCTION

Apart from briefly exemplifying the idea of two-way tables in Section 2.3.1, the summary techniques we have so far considered have been **univariate**, i.e. relating to a single variable. However, most studies are essentially **multivariate** in character. This is partly because having 'captured' a sample unit (e.g. by persuading a potential respondent to be interviewed or by delineating a field plot, sowing a crop therein and applying the combination of fertilizers required by the study), it is cost-effective to ask a moderately large number of questions or to measure several attributes of possible interest. But, quite apart from this, it is fundamental in scientific research to explore relationships of cause and effect and such investigations must first consider whether or not there exist associations among random variables. At the simplest level, we need to be able to measure by an appropriate sample statistic the degree of association between two random variables – the **bivariate** case.

For instance, if we could demonstrate an association between whether or not women took a certain drug during pregnancy and whether or not their babies were born with some specified defect then we might be a step along the road to showing that one caused the other. We would be considering the relationship between the two qualitative random variables 'presence or absence of drug' and 'presence or absence of defect'.

Likewise, if we could show that trees grown from seed collected at several wild locations (i.e. provenances) have different average growth rates then we have demonstrated an association between the qualitative variable 'provenance' and the quantitative variable 'growth rate'. Even a simple statement such as 'Women live longer than men' is making a comment about the relationship that exists between the qualitative variable 'sex' and the quantitative variable 'age at death'.

Although these instances involving at least one qualitative random variable are perfectly valid examples of association, the concept is at its most powerful when applied to two quantitative variables. In particular, we consider briefly in the next section how we could measure the degree of association that might exist between two continuous random variables, x and y, e.g. height and weight of 25-year old men. How can we summarize the extent to which tall men tend to be heavy and short men tend to be light?

4.2 CORRELATION

We can always represent a continuous bivariate sample of n in the form of a **scatter diagram** (sometimes abbreviated to scattergram), in which the y-values are plotted against the x-values. Suppose the scatter diagram looks something like that in Fig. 4.1(a). Then, whenever x is greater than $\bar{x}$, y tends to be greater than $\bar{y}$ so that $(x - \bar{x})(y - \bar{y})$ is positive; whenever x is below $\bar{x}$, y tends to be below $\bar{y}$ so that, again, $(x - \bar{x})(y - \bar{y})$ is positive. Thus, for such data, we would expect a large positive value for S_{xy}, the sum of these products over the n sample pairs. On the other hand, for Fig. 4.1(b), whenever x is above $\bar{x}$, y tends to be below $\bar{y}$, and vice versa, so that $(x - \bar{x})(y - \bar{y})$ tends to be negative, giving a large negative value for S_{xy}. With the pattern as in Fig. 4.1(c), when x is above $\bar{x}$, y is just about as often below as above $\bar{y}$, and similarly when x is below $\bar{x}$, so that S_{xy} tends to be near zero.

Thus the sample **covariance**

$$\text{cov}(x, y) = S_{xy}/(n - 1)$$

is an obvious candidate for measuring association between x and y in the sample. (The reader will correctly guess that $S_{xy} = \sum(x - \bar{x})(y - \bar{y})$ is best computed as $\sum xy - n\bar{x}\bar{y}$.)

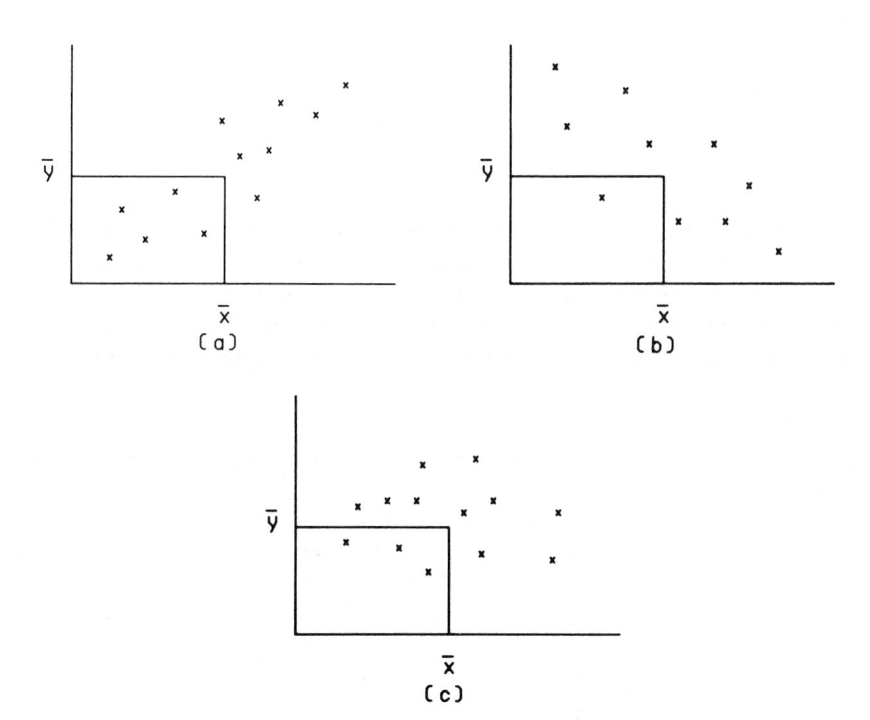

Fig. 4.1 Some possible patterns of scatter diagrams. (a) $r_{xy} > 0$; (b) $r_{xy} < 0$; (c) $r_{xy} = 0$.

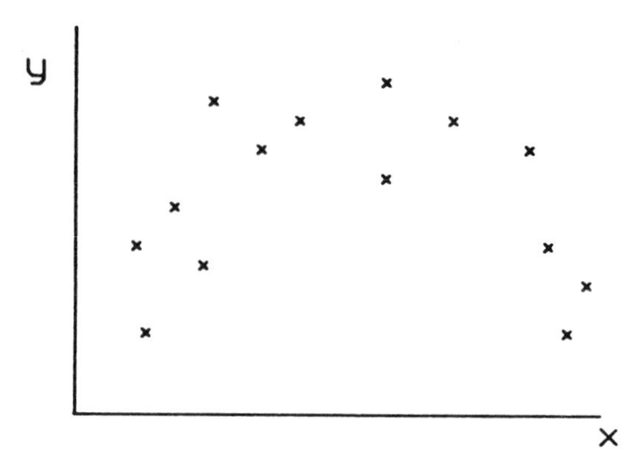

Fig. 4.2 Scattergram for two variables that are related but not linearly

However, the covariance suffers from the drawback that its value depends on the units of measurement of x and y, e.g. if we measure x in inches rather than feet, the covariance is multiplied by 12.

We can get round this by dividing by the sample standard deviations of x and y to give the sample **correlation** (coefficient)

$$r_{xy} = \frac{\sum(x-\bar{x})(y-\bar{y})}{\sqrt{\sum(x-\bar{x})^2 \times \sum(y-\bar{y})^2}} = \frac{S_{xy}}{\sqrt{S_{xx} \times S_{yy}}}$$

If x and y are measured respectively in feet and lbs., say, then $\text{cov}(x, y)$ is in (ft lbs). But the denominator of r_{xy} is also in $\sqrt{(\text{ft}^2\ \text{lbs}^2)} = \text{ft lbs}$, so that r_{xy} is dimensionless and would have the same value for measurements in centimetres and kilograms.

Notice that:

1. Covariance and correlation are sample statistics that deal only with *linear* association; a scatter diagram such as that of Fig. 4.2 shows a clear relationship between x and y, yet r_{xy} could well be zero.
2. The maximum value of r_{xy} is $+1.0$, when x varies in perfect unison with y, and the minimum value is -1.0, when x varies in perfect unison with $(-y)$.

Example 4.1

Suppose we have the mathematics marks and statistics marks for five students as follows:

Maths % (x)	Stats % (y)
64	70
72	68
60	45
80	77
42	47

A scattergram of these gives us no reason to suppose that any relationship is other than linear. Hence we can compute the sample correlation as follows:

$$\bar{x} = 63.6 \qquad \bar{y} = 61.4$$
$$\sum x^2 = 21\,044 \qquad \sum y^2 = 19\,687$$
$$\therefore S_{xx} = 819.2 \qquad \therefore S_{yy} = 837.2$$
$$\sum xy = 20\,210$$
$$\therefore S_{xy} = 684.8$$
$$\therefore r_{xy} = \frac{684.8}{\sqrt{819.2 \times 837.2}} = 0.827$$

(Correlations are usually quoted to three decimal places.)

It should be noted that the importance attached to any particular magnitude of correlation depends on the sample size. A discussion of this in detail needs a fairly full exposition of the inferential methods presented only in outline in the next chapter. Suffice it to say that, for only five students, a correlation between mathematics and statistics marks of only 0.827 is not excitingly high; on the other hand, for a sample of 40, a correlation of only 0.3 in magnitude would be moderately interesting.

4.3 SIMPLE LINEAR REGRESSION

The correlation coefficient tells us about the *strength* of a linear relationship between random variables x and y. But we usually want to be more explicit about the form of the relationship so that we can use it to *predict*, i.e. to find an estimate of y when we know only the value of x.

We can think of the relationship as a straight line – the **regression line**– through the scatter diagram (Fig. 4.3); points (x, y) on the line satisfy the equation

$$y = \alpha + \beta x$$

Notice that we are not saying that for a given value of x, the observed value of y must be $\alpha + \beta x$. The regression line tells us rather that the population *average* of y is $\alpha + \beta x$, i.e. the line is such that when x is zero, the average value of y is α and this average increases by β for every unit increase in the value of x; when β is negative, the line slopes downwards from left to right. We (usually) also require that the variability of y about its mean is constant over different values of x.

The value of α is the **intercept** or **constant** and β is the **slope** or **regression coefficient**.

To make this more concrete, suppose we are talking about heights in centimetres (x) and weights in kilograms (y) of (say) 25-year-old men and consider only the subpopulation that are 160 cm tall; the weights of men in this population are to be distributed (with a variance the same as that of the subpopulations

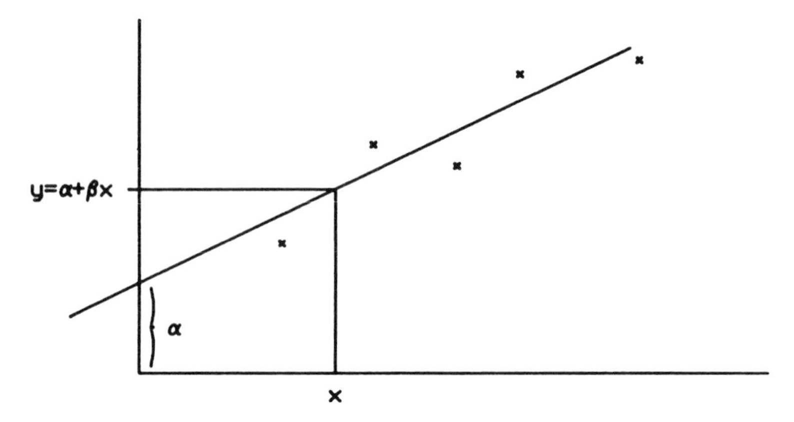

Fig. 4.3 Scatter diagram and regression line

defined by any other height) about a mean $\alpha + \beta \times 160$ (see Fig. 4.4) where

$$\alpha \text{ is in kg}$$
$$\beta \text{ is in kg/cm}$$

Notice that α, β are *not* dimensionless; if we measured in inches and lbs, α and β would change.

Notice, also, that the regression relationship of *y on x* does not place the variables on an equal footing (as does the correlation coefficient). We are predicting *y from x*, i.e. weight *from* height. The regression of height on weight might also be interesting but the intercept and slope would be completely

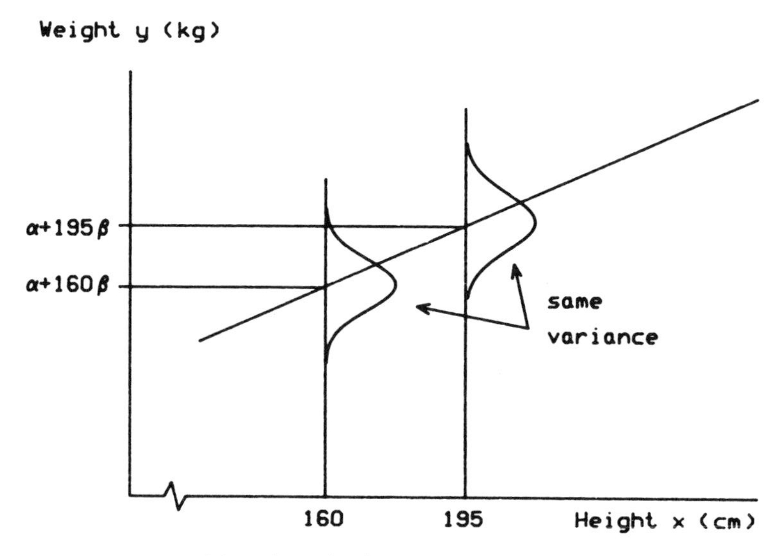

Fig. 4.4 Variability of weight about the average for a given height

different numerically and measured in units of cm and cm/kg respectively. Sometimes regression methods are used to provide a *calibration* where a crude measuring technique (x) is employed in place of a difficult or expensive but accurate one (y); for example one might want to predict the timber volume (y) of a tree from its breast-height diameter (x). In such a case, reversing the roles of x and y would not make sense.

Now let us face the fact that α and β are population constants whose value we can never know. What sample statistics, $\hat{\alpha}$ and $\hat{\beta}$, should represent them?

We consider (Fig. 4.5) the extent to which each observed y *deviates* from its predicted value $\hat{y} = \hat{\alpha} + \hat{\beta}x$. The quantity $y - \hat{y}$ is called the **residual** for that observation. An overall measure of closeness of the line to the data points will be the sum of the squared residuals

$$\text{RSS} = \sum(y - \hat{\alpha} - \hat{\beta}x)^2$$

and it is obviously sensible to find the $\hat{\alpha}$ and $\hat{\beta}$ that minimize RSS. It turns out that the appropriate formulae are

$$\hat{\beta} = S_{xy}/S_{xx}$$

and

$$\hat{\alpha} = \bar{y} - \hat{\beta}\bar{x}$$

so that

$$\text{RSS} = S_{yy} - S_{xy}^2/S_{xx}$$

The basic variability of y about the regression line is estimated by RSS/$(n-2)$, called the **residual mean square** (RMS); this provides a measure of the 'goodness of fit' of the regression relationship.

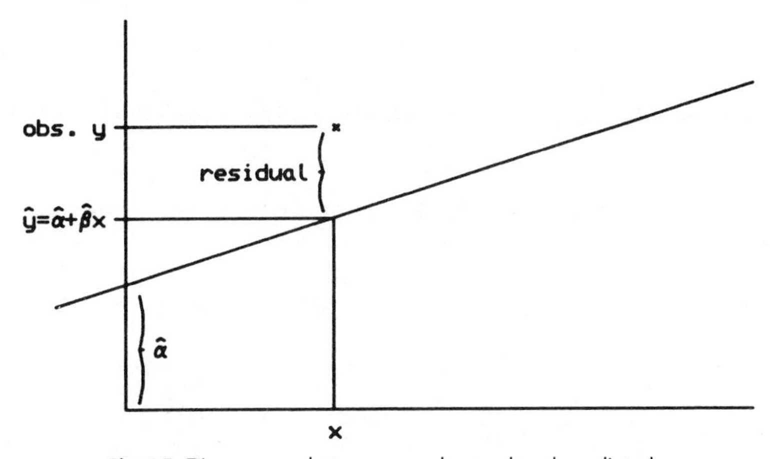

Fig. 4.5 Discrepancy between an observed and predicted y

An aside

Those of a mathematically nervous disposition should perhaps ignore the following paragraph of explanation of these formulae. Obviously, calculus methods can be used to minimize RSS fairly straightforwardly. Alternatively, symmetry suggests that the sum of the residuals should be zero, i.e.

$$\sum(y - \hat{\alpha} - \hat{\beta}x) = 0$$

which means that $\hat{\alpha} = \bar{y} - \hat{\beta}\bar{x}$
Thus

$$\begin{aligned}
\text{RSS} &= \sum(y - \hat{\alpha} - \hat{\beta}x)^2 \\
&= \sum[(y - \bar{y}) - \hat{\beta}(x - \bar{x})]^2 \\
&= S_{yy} - 2\hat{\beta}S_{xy} + \hat{\beta}^2 S_{xx} \\
&= \left(S_{yy} - \frac{S_{xy}^2}{S_{xx}}\right) + \frac{S_{xy}^2}{S_{xx}} - 2\hat{\beta}S_{xy} + \hat{\beta}^2 S_{xx} \\
&= \left(S_{yy} - \frac{S_{xy}^2}{S_{xx}}\right) + (S_{xy} - \hat{\beta}S_{xx})^2/S_{xx}
\end{aligned}$$

Hence RSS must be minimized when the second term is zero, i.e. when $\hat{\beta} = S_{xy}/S_{xx}$ at which point, RSS $= S_{yy} - (S_{xy}^2/S_{xx})$. (Notice that RSS $= 0$ if correlation $= \pm 1$ which makes sense.) This completes the explanation.

Suppose that, using the data in Example 4.1, we wished to predict the statistics mark, y, for a student with a mathematics mark of x. From the previous calculations, we immediately have

$$\hat{\beta} = \frac{S_{xy}}{S_{xx}} = \frac{684.8}{819.2} = 0.8359$$

$$\hat{\alpha} = \bar{y} - \hat{\beta}\bar{x} = 61.4 - 0.8359 \times 63.6 = 8.23$$

$\therefore$ on average, statistics % $= 8.23 + 0.8359 \times$ mathematics %
Thus, a predicted statistics mark for a student with a mathematics mark of 55% is

$$\hat{y} = 8.23 + 0.8359 \times 55 = 54.20$$

We can do this for the students in the sample and hence calculate residuals as follows:

Maths % (x)	Stats % (y)	Predicted ($\hat{y}$)	Residual ($y - \hat{y}$)
64	70	61.73	8.27
72	68	68.42	−0.42
60	45	58.39	−13.39
80	77	75.11	1.89
42	47	43.35	3.65
			$\sum = 0$ (approx.)

Note the following:

1. the sum of the residuals is zero (allowing for rounding)
2. the line passes through the point $(\bar{x}, \bar{y})$ because, when $x = \bar{x}$, $y = (\bar{y} - \hat{\beta}\bar{x}) + \hat{\beta}\bar{x} = \bar{y}$
3. $\hat{\alpha}$ is in units of 'statistics marks', $\hat{\beta}$ is in units of 'statistics marks per mathematics mark'
4. the sum of the squared residuals is 264.756 but this is more easily computed as

$$\text{RSS} = S_{yy} - \frac{S_{xy}^2}{S_{xx}} = 837.2 - \frac{684.8^2}{819.2} = 264.75$$

(The discrepancy is due to rounding of the residuals.)

Notice that the asymmetry between x and y means that x is not necessarily a random variable. In cases where x and y are determined from a random sample of n units, then x will indeed have a parent distribution though its form is irrelevant. But, in an experimental situation, the levels of x can be chosen by the experimenter and it is clearly only the y-values that vary randomly.

The relationship between x and y is often called the **model**; we have been considering the simple linear model. The ideas can be extended to allow consideration of more than one x-variable (**multiple regression**) but interpretation of such models is very difficult unless the levels of the x-variables are carefully chosen. Regression methods can also be applied to *non-linear* models when the background to the problem or inspection of the scattergram suggest a linearizing transformation.

Example 4.2
Consider the following concocted data:

x	y
1	0.50
2	1.12
3	1.60
4	2.00
5	2.35

Figure 4.6 shows the scatter diagram.

The linear regression is

$$\hat{y} = 0.140 + 0.458x \qquad \text{with RMS} = 0.009\,43$$

but the fictional experimenter whose data these are now tells us that his theory suggests y is related to $\sqrt{x}$ rather than x; such a model is certainly consistent with the slight curvature of the scatter plot. Transforming x to $x' = \sqrt{x}$ and regressing

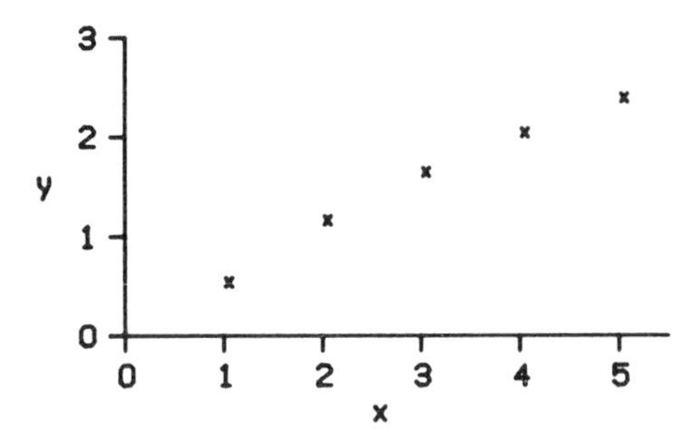

Fig. 4.6 Scattergram for the data of Example 4.2

y on x' now gives

$$\hat{y} = -0.997 + 1.498\sqrt{x} \qquad \text{with RMS} = 0.000\,056\,2$$

The much better fit of the second model (lower residual mean square) seems to support the experimenter's contention. (In fact, the y-values were constructed by rounding $-1.0 + 1.5\sqrt{x}$ to two decimal places.)

Correlation and regression calculations are time-consuming for even moderate amounts of data but, luckily, the smallest microcomputer (and even some pocket calculators) can handle the problem fairly straightforwardly. In practice, the examination of regression residuals is very important both to detect aberrant observations (**outliers**) and to confirm assumptions about the model. For example, a scattergram of the residuals plotted against x may show up inconstancy of variance.

4.3.1 Standard error of prediction

We have seen how to make a prediction of y for any given value, x_0, of x, i.e.

$$\hat{y}_0 = \hat{\alpha} + \hat{\beta}x_0$$

Do not forget, however, that $\hat{y}_0$ is a random variable – its value would vary over different selections of the original n sample members. Hence, $\hat{y}_0$ has an associated standard error given by the somewhat revolting looking formula

$$\sqrt{\text{RMS}\left(\frac{1}{n} + \frac{(x_0 - \bar{x})^2}{S_{xx}}\right)}$$

Thus, for the mathematics/statistics marks of Example 4.1,

$$\text{SE}(\hat{y}_0) = \sqrt{\frac{264.75}{3}\left(\frac{1}{5} + \frac{(x_0 - 63.6)^2}{819.2}\right)}$$

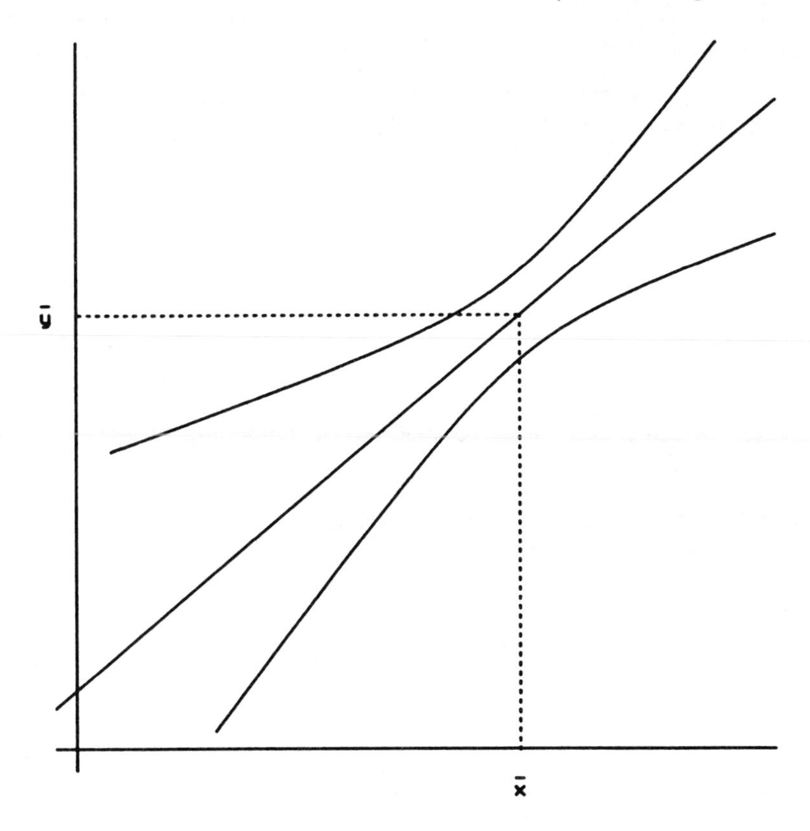

Fig. 4.7 Curves showing the uncertainty of prediction of y as x varies

$$\text{i.e. if } x_0 = 60, \text{SE}(\hat{y}_0) = \sqrt{\frac{264.75}{3}\left(\frac{1}{5} + \frac{3.6^2}{819.2}\right)} = 4.36$$

$$\text{if } x_0 = 80, \text{SE}(\hat{y}_0) = \sqrt{\frac{264.75}{3}\left(\frac{1}{5} + \frac{16.4^2}{819.2}\right)} = 6.83$$

Notice that because of the term $(x_0 - \bar{x})^2$, the standard error of prediction gets larger the further we move away from the mean. Figure 4.7 shows this. But, the prediction for a value of x *outside* the range of x-values used to calculate the regression line will be liable to considerable error not only because of this but also because the relationship of x and y may not be of the same form outside the area investigated. The estimation of y for x outside the range of the sample is called **extrapolation**. We should extrapolate only with the utmost caution particularly when x represents time and we are attempting to predict the future.

4.4 A REAL ASSOCIATION?

Suppose that sample data have provided some evidence that two variables, x and y (quantitative or qualitative) are associated. Then we can ask: 'Is this apparent association "real" or can it be explained by a common association with a third variable, z?'

Example 4.3
Suppose a social scientist is looking at sport participation among 15-year-old schoolboys; for each of 500 boys in Catholic (C) schools and 500 boys in non-Catholic (NC) schools, he records:

(i) Social class – middle or working.
(ii) A teacher categorization as 'competitive' or 'non-competitive'.
(iii) Time to run 600 metres (secs).

He looks first at the running times and finds:

	NC	C	Overall
Average time	122	126	124
n	500	500	1000

There appears to be a 'religious' difference of 4 seconds, i.e. an association between type of school attended and running ability. Of course, this may well be due to the vagaries of sampling, but let us assume in this chapter that the difference is a real one.

Suppose we further subdivide by the social class grouping. We could find any of the following patterns (and others):

(a) A genuine C/NC difference in both middle and working classes, e.g.

	Middle			Working		
	NC	C	Overall	NC	C	Overall
Average time	122	126	123.3	122	126	125
n	400	200	600	100	300	400

(b) No C/NC difference at all – instead a speed difference between social classes, e.g.

	Middle			Working		
	NC	C	Overall	NC	C	Overall
Average time	120	120	120	130	130	130
n	400	200	600	100	300	400

(c) An interaction between type of school and social class, i.e. the speed differential between the two types of school is different in the two social classes. For example, in the extreme case, we might have a C/NC difference *only* in the working class.

	Middle			Working		
	NC	C	Overall	NC	C	Overall
Average time	123	123	123	118	128	125.5
n	400	200	600	100	300	400

Thus we can see how an apparent association between a quantitative variable (time) and a qualitative variable (C/NC school) can be examined in relation to a third variable (social class).

Suppose our social scientist finds an apparent relationship between the competitiveness categorization and type of school – two qualitative random variables, e.g. the two-way table might give numbers such as:

	NC	C
compet.	180	140
non-compet.	320	360
	500	500
compet. % =	36%	28%

As before, we can enquire whether this could be explained by social class differences. We might find patterns such as the following:

(a) A consistent association in both social class groups, e.g.

	Middle			Working		
	NC	C	Overall	NC	C	Overall
compet.	144	56	200	36	84	120
non-compet.	256	144	400	64	216	280
	400	200	600	100	300	400
compet. % =	36%	28%	33%	36%	28%	30%

(b) The apparent association is an artefact due to common association with social class, e.g.

	Middle			Working		
	NC	C	Overall	NC	C	Overall
compet.	160	80	240	20	60	80
non-compet.	240	120	360	80	240	320
	400	200	600	100	300	400
compet. % =	40%	40%	40%	20%	20%	20%

(c) An even greater association but in only one of the social class groups, e.g.

	Middle			Working		
	NC	C	Overall	NC	C	Overall
compet.	140	70	210	40	70	110
non-compet.	260	130	390	60	230	290
	400	200	600	100	300	400
compet. % =	35%	35%	35%	40%	23.3%	27.5%

These examples have shown how relationships between two random variables can be studied by considering their joint association with other variables. (Similar ideas apply when both x and y are quantitative.) There can be no rules about how this elucidation should proceed when many variables are involved. Where a statistician works constantly with data from a certain application area (such a person is sometimes called a topic statistician), he or she develops a 'feel' for what to do; otherwise extensive discussion with experts in that field is required.

4.5 CAUSE AND EFFECT

It is a great fallacy to assume that, because variables x and y are associated, x is the *cause* of an *effect* y, or vice versa.

Certainly, an association must be demonstrated, e.g. if smoking during pregnancy *causes* congenital abnormalities, smoking mothers-to-be would have to have a higher abnormality rate in their babies than non-smokers (a cause-to-effect argument) or there would have to be a higher proportion of smokers mothers among sufferers from congenital abnormalities than for non-sufferers (an effect-to-cause argument).

If y precedes x in time, then x cannot cause y, e.g. if the appearance of liver disease preceded heavy alcohol consumption the disease could not be the result of the drinking. But perhaps the symptoms follow so long after onset of the disease that it is not clear that desire to drink is not stimulated *by* liver disease! However unlikely, it is a point that must be considered.

Just as association can sometimes be elucidated by consideration of further variables, so the mechanisms of cause and effect require consideration of 'outside' factors. Thus the association between grey hair and high blood pressure does not mean that one causes the other but simply that both are usually the result of ageing. Indeed the passage of time is an explanation of many apparent relationships, e.g. between sales of television sets and the divorce rate over the last thirty years. Sometimes a cause-and-effect linkage is indirect. Thus a declining beech tree population might bring about an associated fall in the bird population but the connection is via the reduced food supply due to the decreasing insect habitat.

SUMMARY

Measurement of the association between variables is a fundamental component of statistical investigation. The correlation between two quantitative random variables quantifies association in a way that does not depend on units of measurement. Regression methods, on the other hand, aim at producing an equation to predict one variable from another. However, any apparent relationship may be explicable by reference to other variables and certainly does not of itself imply cause and effect.

EXERCISES

(Readers with some computing skills may choose to employ electronic assistance in some of these calculations. The time saved thereby should be dedicated to thinking about the meaning of the results.)

1. Refer back to Exercise 4 of Chapter 2 on Becker muscular dystrophy. Regard the SCK level as y and age as x.
   (a) The ages of the normal women were known and, together with the SCK levels gave the following summary statistics:

$$\sum x = 5050.5 \qquad \sum y = 3300.4$$
$$\sum x^2 = 257\,019.25 \qquad \sum y^2 = 107\,470.4$$
$$\sum xy = 137\,731.2$$

   Find the correlation between age and SCK and the regression equation linking age and SCK level. Explain what the equation tells you. What would be the predicted SCK levels and standard errors for women aged 20 and 40?
   (b) The ages corresponding to SCK levels for the 30 known carriers were recorded as:

70	58	5.5	61	14	50	79	47	54	31
74	77	32	42	64	35	40	65	43	45
72	69	44	56	50	64	74	71	39	62

   Plot the scattergram of SCK levels against age. Repeat the calculations in (a) for these carriers.
   (c) Plot both regression lines on the scattergram for carriers and comment on the usefulness of SCK levels as a means of detecting carriers of Becker muscular dystrophy.

2. An ecologist presents you with data on soil moisture content and vegetation cover for a sample of 20 plots laid out on a mountainside see Table E4.1.

Table E4.1 Soil moisture content (x) and vegetation cover (y) for land plots

Moisture (x)	Vegetation (y)	Moisture (x)	Vegetation (y)
72.1	8.2	74.6	11.9
67.6	3.3	77.4	22.4
70.9	2.2	79.1	21.2
69.2	2.6	77.7	10.0
47.5	1.0	78.1	12.5
60.7	4.1	72.8	9.5
38.8	5.7	78.3	19.2
59.0	1.7	75.2	11.0
40.5	5.0	72.6	7.4
55.1	3.1	75.1	9.0

(a) Draw the scatter diagram.
The relationship does not appear to be linear. (In fact, the residual mean square for linear regression is 26.82.)

(b) You are just about to consider a model such as

$$y = \alpha + \beta(x - 55)^2$$

(for which the residual mean square is reduced to 11.19), when the ecologist remarks that the ten plots listed on the left above were all at around 500 m altitude whereas the right-hand ten plots were at 700 m. Compute the separate regressions for the plots at the two altitudes and add them to your scattergram.

(c) Express your findings using the word 'interaction'. Can you suggest any biological explanation?

3. Chapman (1980) considers possible relationships between the weights (in tons) of particles emitted (during 1973) from 31 petrochemical plants in Texas and a measure of the size of the establishments, and the data are given in Table E4.2.

Table E4.2 Emission of particles from Texas petrochemical plants

Plant no.	Size	Particle wt.	Plant no.	Size	Particle wt.
1	2	23.1	17	2	8.3
2	5	10.1	18	9	15.6
3	8	2.7	19	9	29.2
4	3	24.7	20	16	36.8
5	8	22.1	21	1	2.7
6	3	14.0	22	2	37.0
7	39	270.2	23	3	0.6
8	10	5.1	24	22	75.8
9	1	23.5	25	4	1.9
10	11	44.1	26	12	59.4
11	2	49.8	27	1	9.0
12	6	28.7	28	5	45.9
13	3	72.5	29	22	42.5
14	4	11.5	30	15	68.6
15	5	22.8	31	10	17.3
16	4	32.0			

(a) Plot particle emission versus plant size incorporating the linear regression line of emission on size in your diagram.

(b) Confirm that plants 7, 13 and 29 are particularly badly fitted by the model, accounting together for about 57% of the sum of squared residuals. Compute a few other residuals for comparison.

(c) Plant 7 is clearly so large that it has a very considerable influence on the estimated regression line – it is said to have high **leverage**. But maybe the true regression relationship is quadratic, i.e.

$$\text{particle emission} = \alpha + \beta \times \text{size}^2$$

Investigate this.

(d) Suppose that it could be shown that the mix of compounds produced at plant 7 made it atypical so that it should validly be excluded from the analysis. Re-compute the linear regression line and add it to your diagram.

(e) Compare the residual mean squares for the three relationships you have produced and comment.

(Chapman, K. (1980) Petrochemicals and pollution in Texas and Louisiana. Internal report, Department of Geography, University of Aberdeen.)

4. (a) Kucerova *et al.* (1972) report on chromosomal damage resulting from irradiation of blood samples by X-rays. At each radiation dose, 1500 cells were examined, the numbers of damaged chromosomes being as follows:

Radiation dose (rads)	0	5	10	15	30	50
Damaged chromosomes per 1500 cells	2	2	2	5	26	31

Plot damage per cell against radiation doseage and add the linear regression line of damage per cell on dose. Does the linear relationship seem reasonable?

(b) Adams (1970) gives similar data for higher doses as follows, 2000 cells being scored at each dose:

Radiation dose (rads)	0	75	150	300	400
Damaged chromosomes per 2000 cells	0	250	770	2150	3710

Again, plot damage per cell versus dose. Do these data appear consistent with the previous linear regression relationship?

By looking at the residual mean squares, compare models for Adam's data that relate damage per cell to

(i) dose

(ii) squared dose.

(c) Discuss the consistency of the two experiments. Combine the two sets of data to give a relationship predicting damage per cell for a given X-ray dose.

(There is a minor technical problem with these analyses: the y-variable is actually discrete and, in particular, the variability of chromosome damage is almost certainly not constant over radiation dose. However, a more valid analysis – beyond the scope of this book – amends the basic findings very little.)

(Adams, A.C. (1970) The response of chromosomes of human peripheral blood lymphocytes to X-rays. PhD Thesis, University of Aberdeen.

Kucerova, M., Anderson, A.J.B., Buckton, K.E. and Evans, H.J. (1972) X-ray-induced chromosome aberrations in human peripheral blood leucocytes: the response to low levels of exposure *in vitro*. *Int. J. Radiat. Biol.*, **21**, 389–396.)

5. In a study of the consumption of leaves by earthworms, thirty metre-square wire-mesh

cages were randomly positioned in an orchard in December and, under each, the ground was raked and a standard weight of leaves scattered. The average soil temperatures at depths of 10 cm and 20 cm in the middle of the orchard and the total weights of leaves consumed during each of the next 15 weeks were as in Table E4.3.

Table E4.3 Weight of leaves consumed by earthworms, and soil temperature

Weight (g)	Temp (°C) at 20 cm	Temp (°C) at 10 cm
197	1.6	−0.9
268	5.6	0.6
238	6.1	1.7
219	1.8	−1.7
252	4.8	0.8
262	4.6	−0.6
204	2.0	−0.1
185	−1.2	−3.9
252	3.3	−1.1
295	0.7	−6.8
267	2.5	−3.8
285	5.6	−0.3
258	6.3	1.6
309	6.7	0.6
275	6.1	1.2

(a) One would expect the temperatures at the two depths to be highly correlated. Check this. It is a fact that the sum and difference of the variates having roughly the same variance tend to be uncorrelated. Check this for these temperatures. (Those with moderate algebraic skill will see how to do this with very little extra calculation. Note also that sums and differences of variables are sensible entities only if the variables are **congeneric**, i.e. measure the same kind of thing.)

(b) We want to relate weight of leaves consumed to the underground temperatures. It *may* be that temperatures at specific depths such as 10 cm or 20 cm are important (the residual mean squares for linear regressions of leaf weight on these temperatures turn out to be 896.9 and 1419.3 respectively). But overall soil temperature and/or the temperature gradient are biologically more likely to influence the worms' appetite, i.e. the sum and difference of temperatures that we have already looked at. Investigate this.

(These analyses are very slightly suspect in that the n observation pairs should be totally unrelated, which may not be true of weeks in a sequence. Biological and statistical judgement suggest the difficulty can be ignored here, but it should be borne in mind particularly when observations are made in such a time sequence.)

6. Examine each of the following statements and the conclusion drawn from it. Assuming

that the statement is true, present carefully reasoned arguments against the validity of the conclusion.

(a) A survey indicated that three out of four Scots had visited their doctors during the preceding month whereas in an English survey the year before, the corresponding figure had been four out of five. This proves the Scots are healthier than the English.
(b) After a campaign of gassing badgers, the proportions of cattle and of badgers suffering from tuberculosis both fell. This shows that badgers spread tuberculosis to cattle.
(c) In a survey of old-age pensioners, it was found that the average weight of women aged 60 was 66.2 kg whereas the corresponding figure for women aged 70 was 64.1 kg. Hence women lose weight as they grow older.
(d) In the USA, a surprisingly high number of general practitioners are of Jewish descent. Thus we conclude that medical skill is inherited.
(e) There is a strong correlation between the amount of sugar eaten in a country and the death rate from coronary heart disease. Hence sugar is an important cause of this illness.

7. (An exercise in algebra.)
Why are the degrees of freedom (the divisor) for the residual mean square equal to $(n-2)$? For much the same reason as there are $(n-1)$ degrees of freedom in the calculation of the sample variance (see Section 3.3); once $(n-2)$ residuals are given, the remaining 2 are known.
Suppose we have

x	y
1	4
2	lost
3	lost

Before loss of the y-values, the regression line was found to be

$$y = -3 + 6x$$

Find the missing y-values and hence their residuals.

5
Testing hypotheses

5.1 INTRODUCTION

The material presented in the last three chapters can be used in the description of samples but often we need to go further. A great many statistical techniques are directed towards inferring the nature of a population when only a sample is available. Somewhat paradoxically, it turns out that we can best approach this problem from the opposite direction – by investigating the nature of samples that could arise from a given population. To do this, we need to introduce the concept of probability.

5.2 PROBABILITY

The **probability** that a randomly selected unit of a population lies in a given category is simply the proportion of the population that belongs to that category, that is to say the **relative frequency** of the category. For instance, consider again the table of BSc students in Example 2.2 reproduced here.

Male/female breakdown of BSc students at Aberdeen University 1984/85

	Men	Women	Total
BSc (Agriculture)	107	48	155
BSc (Engineering)	304	30	334
BSc (Forestry)	93	7	100
BSc (Pure Science)	636	635	1271
Total	1140	720	1860

Source: University of Aberdeen *Annual Report* 1985

The probability that a randomly chosen science student is aiming at an engineering degree is 334/1860 ($=0.1796$); the probability of being a female science student is 720/1860 ($=0.3871$).

Certain properties of probability are immediately obvious:

1. Probabilities lie between 0.0 and 1.0 inclusive. Zero corresponds to an impossible category, i.e. one that contains no members of the population, and

1.0 is the probability of a category to which every member of the population must belong.

2. The probability of not being in a given category equals 1.0 minus the probability of being in the category, e.g. probability that a randomly selected science student is male

$$= \frac{1140}{1860} = \frac{1860 - 720}{1860} = 1 - \frac{720}{1860}$$

$$= 1 - \text{probability student is female, i.e. not male}$$

3. As an extension to 2, when categories are mutually exclusive and exhaustive, that is, each unit belongs to one and only one category, then the sum of the probabilities of the different categories is 1.0. Thus

$$\text{Prob(agri.)} + \text{Prob(eng.)} + \text{Prob(for.)} + \text{Prob(pure sci.)}$$

$$= \frac{155}{1860} + \frac{334}{1860} + \frac{100}{1860} + \frac{1271}{1860} = \frac{1860}{1860} = 1.0$$

5.2.1 Independence

By **conditional probability** we mean, for example, the probability that a randomly chosen science student is male *given that* he is a forester, i.e. if we are told that our random student is a forester, we must naturally modify the probability that he is male from 1140/1860 to 93/100 (= 0.9300). This leads to the important idea of **independence**: two random variables are said to be independent if the value taken by one tells us nothing about the probable value of the other. (This is clearly linked with the ideas of association introduced in Chapter 4.)

We have just seen that sex and type of degree are *not* independent (which accords with common experience). If sex and degree *were* to be independent, the population of science students would have to have a breakdown such as:

	Men	Women	Total
BSc (Agriculture)	95	60	155
BSc (Engineering)	190	120	312
BSc (Forestry)	76	48	124
BSc (Pure Science)	779	492	1271
Total	1140	720	1860

If that were the case,

$$95/155 = 190/310 = 76/124 = 779/1271 = 1140/1860$$

and knowledge of degree would tell us nothing about the probable sex of a student. Likewise, knowledge of a randomly selected student's sex would tell us nothing about their probable degree intention.

5.3 PROBABILITY DISTRIBUTIONS

The definition of probability based on relative frequency requires a little clarification when applied to infinite populations; the top and bottom lines of the relative frequency are now infinite. Nevertheless we can still think of the probability associated with any category as being the proportion of the population in that category. This will allow us to introduce the concept of a theoretical **probability distribution** where the probability of any category can be calculated from theoretical considerations but for which the population can never be complete in reality. For example, in the tossing of an unbiased coin, we can say of the distribution of probability over the heads/tails categories that both probabilities must be 1/2 even though at any point in time we can have physically generated only an infinitesimal fraction of the hypothetical infinite population of coin tosses. This is because heads and tails are equally likely and

$$\text{Prob(heads)} + \text{Prob(tails)} = 1.0$$

Likewise for a true six-sided die, the probability of any given face landing uppermost is 1/6.

Now suppose we throw the die twice and count the number, r, of fours (say) arising from the 36 equally likely outcomes; thus r is a sample statistic that can take the values 0, 1 or 2.

Number of fours

		Score on second throw					
		1	2	3	4	5	6
Score on first throw	1	0	0	0	1	0	0
	2	0	0	0	1	0	0
	3	0	0	0	1	0	0
	4	1	1	1	2	1	1
	5	0	0	0	1	0	0
	6	0	0	0	1	0	0

Only one of 36 possibilities gives rise to a double four, so

$$\text{Prob } (r = 2 \text{ fours}) = 1/36$$

There are 10 cases where exactly one four arises, so

$$\text{Prob } (r = 1 \text{ four}) = 10/36$$

Table 5.1 Binomial distributions for the probability of r out of n being in a category whose probability is p.

n	r	$p = 0.1$	0.143 = 1/7	0.167 = 1/6	0.2	0.25	0.3	0.333 = 1/3	0.4	0.5
4	0	0.6561	0.5394	0.4815	0.4096	0.3164	0.2401	0.1979	0.1296	0.0625
	1	0.2916	0.3600	0.3861	0.4096	0.4219	0.4116	0.3953	0.3456	0.2500
	2	0.0486	0.0901	0.1161	0.1536	0.2109	0.2646	0.2960	0.3456	0.3750
	3	0.0036	0.0100	0.0155	0.0256	0.0469	0.0756	0.0985	0.1536	0.2500
	4	0.0001	0.0004	0.0008	0.0016	0.0039	0.0081	0.0123	0.0256	0.0625
7	0	0.4783	0.3395	0.2783	0.2097	0.1335	0.0824	0.0587	0.0280	0.0078
	1	0.3720	0.3966	0.3906	0.3670	0.3115	0.2471	0.2053	0.1306	0.0547
	2	0.1240	0.1985	0.2349	0.2753	0.3115	0.3177	0.3074	0.2613	0.1641
	3	0.0230	0.0552	0.0785	0.1147	0.1730	0.2269	0.2558	0.2903	0.2734
	4	0.0026	0.0092	0.0157	0.0287	0.0577	0.0972	0.1277	0.1935	0.2734
	5	0.0002	0.0009	0.0019	0.0043	0.0115	0.0250	0.0383	0.0774	0.1641
	6	0.0000	0.0001	0.0001	0.0004	0.0013	0.0036	0.0064	0.0172	0.0547
	7	0.0000	0.0000	0.0000	0.0000	0.0001	0.0002	0.0005	0.0016	0.0078
10	0	0.3487	0.2137	0.1609	0.1074	0.0563	0.0282	0.0174	0.0060	0.0010
	1	0.3874	0.3566	0.3225	0.2684	0.1877	0.1211	0.0870	0.0403	0.0098
	2	0.1937	0.2678	0.2909	0.3020	0.2816	0.2335	0.1955	0.1209	0.0439
	3	0.0574	0.1191	0.1555	0.2013	0.2503	0.2668	0.2603	0.2150	0.1172
	4	0.0112	0.0348	0.0546	0.0881	0.1460	0.2001	0.2274	0.2508	0.2051
	5	0.0015	0.0070	0.0131	0.0264	0.0584	0.1029	0.1362	0.2007	0.2461
	6	0.0001	0.0010	0.0022	0.0055	0.0162	0.0368	0.0567	0.1115	0.2051
	7	0.0000	0.0001	0.0003	0.0008	0.0031	0.0090	0.0162	0.0425	0.1172
	8	0.0000	0.0000	0.0000	0.0001	0.0004	0.0014	0.0030	0.0106	0.0439
	9	0.0000	0.0000	0.0000	0.0000	0.0000	0.0001	0.0003	0.0016	0.0098
	10	0.0000	0.0000	0.0000	0.0000	0.0000	0.0000	0.0000	0.0001	0.0010

30	0	0.0424	0.0098	0.0042	0.0012	0.0002	0.0000	0.0000	0.0000	0.0000
	1	0.1413	0.0489	0.0250	0.0093	0.0018	0.0003	0.0001	0.0000	0.0000
	2	0.2277	0.1182	0.0728	0.0337	0.0086	0.0018	0.0006	0.0000	0.0000
	3	0.2361	0.1841	0.1362	0.0785	0.0269	0.0072	0.0027	0.0003	0.0000
	4	0.1771	0.2073	0.1843	0.1325	0.0604	0.0208	0.0090	0.0012	0.0000
	5	0.1023	0.1799	0.1921	0.1723	0.1047	0.0464	0.0234	0.0041	0.0001
	6	0.0474	0.1251	0.1605	0.1795	0.1455	0.0829	0.0487	0.0115	0.0006
	7	0.0180	0.0716	0.1103	0.1538	0.1662	0.1219	0.0833	0.0263	0.0019
	8	0.0058	0.0343	0.0636	0.1106	0.1593	0.1501	0.1196	0.0505	0.0055
	9	0.0016	0.0140	0.0312	0.0676	0.1298	0.1573	0.1459	0.0823	0.0133
	10	0.0004	0.0049	0.0131	0.0355	0.0909	0.1416	0.1530	0.1152	0.0280
	11	0.0001	0.0015	0.0048	0.0161	0.0551	0.1103	0.1389	0.1396	0.0509
	12	0.0000	0.0004	0.0015	0.0064	0.0291	0.0749	0.1098	0.1474	0.0806
	13	0.0000	0.0001	0.0004	0.0022	0.0134	0.0444	0.0759	0.1360	0.1115
	14	0.0000	0.0000	0.0001	0.0007	0.0054	0.0231	0.0460	0.1101	0.1354
	15	0.0000	0.0000	0.0000	0.0002	0.0019	0.0106	0.0245	0.0783	0.1445
	16	0.0000	0.0000	0.0000	0.0000	0.0006	0.0042	0.0115	0.0489	0.1354
	17	0.0000	0.0000	0.0000	0.0000	0.0002	0.0015	0.0047	0.0269	0.1115
	18	0.0000	0.0000	0.0000	0.0000	0.0000	0.0005	0.0017	0.0129	0.0806
	19	0.0000	0.0000	0.0000	0.0000	0.0000	0.0001	0.0005	0.0054	0.0509
	20	0.0000	0.0000	0.0000	0.0000	0.0000	0.0000	0.0001	0.0020	0.0280
	21	0.0000	0.0000	0.0000	0.0000	0.0000	0.0000	0.0000	0.0006	0.0133
	22	0.0000	0.0000	0.0000	0.0000	0.0000	0.0000	0.0000	0.0002	0.0055
	23	0.0000	0.0000	0.0000	0.0000	0.0000	0.0000	0.0000	0.0000	0.0019
	24	0.0000	0.0000	0.0000	0.0000	0.0000	0.0000	0.0000	0.0000	0.0006
	25	0.0000	0.0000	0.0000	0.0000	0.0000	0.0000	0.0000	0.0000	0.0001
	26	0.0000	0.0000	0.0000	0.0000	0.0000	0.0000	0.0000	0.0000	0.0000
	27	0.0000	0.0000	0.0000	0.0000	0.0000	0.0000	0.0000	0.0000	0.0000
	28	0.0000	0.0000	0.0000	0.0000	0.0000	0.0000	0.0000	0.0000	0.0000
	29	0.0000	0.0000	0.0000	0.0000	0.0000	0.0000	0.0000	0.0000	0.0000
	30	0.0000	0.0000	0.0000	0.0000	0.0000	0.0000	0.0000	0.0000	0.0000

and, similarly,

$$\text{Prob } (r = 0 \text{ fours}) = 25/36$$

Notice, as usual, that

$$\text{Prob } (r = 0 \text{ fours}) + \text{Prob } (r = 1 \text{ four}) + \text{Prob } (r = 2 \text{ fours}) = 1.0$$

What we have done is to produce the probability distribution for the sample statistic that is the count of the number of fours in two throws of a die; conceptually we could obviously produce the corresponding distribution for any number of throws, although the process becomes more and more laborious as the number of throws is increased. Luckily, some mathematics, which need not concern us, makes the calculation quite feasible.

The kind of probability distribution we have studied here is called the **binomial distribution** (for reasons that are not important for the present discussion); it applies to counts, r, of the occurrence of some category of interest in a sample where the probability of that category appearing is constant for all sample members. The values in the table of the probability distribution and the shape of the corresponding histogram are completely determined when the number, n, in the sample and the probability, p, of the category of interest are known. Table 5.1 gives the probabilities for binomial distributions with a few selected values of n and p. (The apparent zero probabilities in the table are, of course, only zeros to the fourth decimal place.) Figures 5.1 and 5.2 compare the histograms for samples of 30 and $p = 1/6$ and $p = 1/2$ respectively; notice how the distribution is skew unless $p = 1/2$.

Many other theoretical probability distributions exist that describe the pattern of probabilities for observations of sample statistics arising from other situations. Examples of four distribution patterns for continuous variates were given in

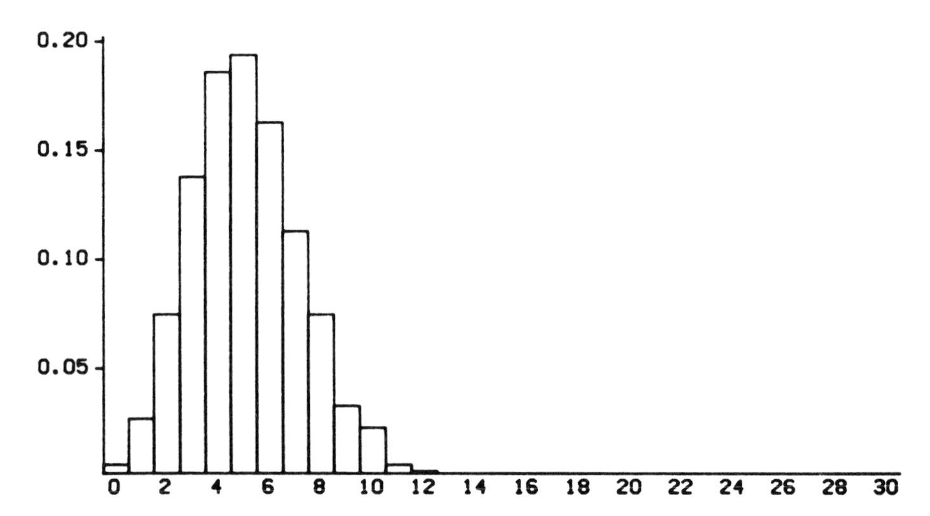

Fig. 5.1 Histogram for binomial distribution for $n = 30$, $p = 1/6$

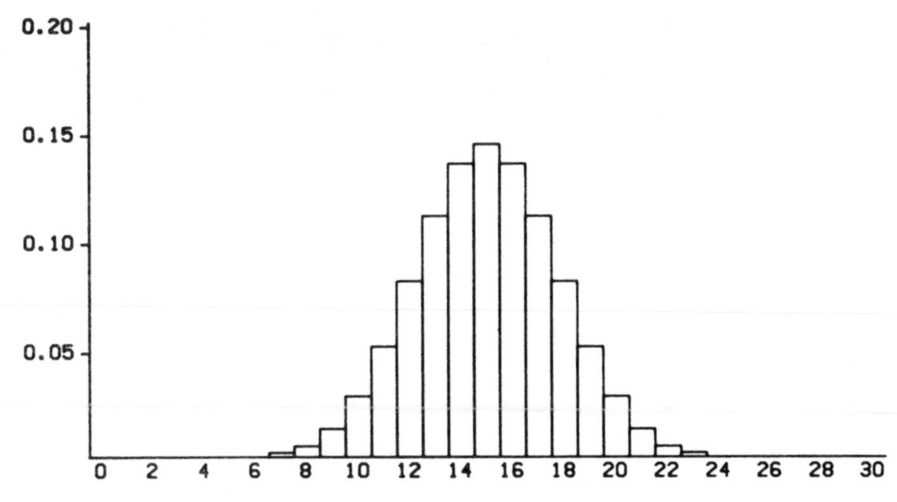

Fig. 5.2 Histogram for binomial distribution for $n = 30$, $p = 1/2$

Fig. 2.8. For experimental studies, the most important is the normal distribution since sample means and the differences between sample means follow this pattern of variability more and more closely the greater the sample size, whatever the pattern of variability of individual observations. For example, the normal distribution could be applied with no greater risk of misrepresentation in studies of 30 or more units. Extensive tables are widely available giving the probabilities of an observation being more than so many standard deviations away from the true mean. But, for most purposes, all that need be remembered is that, *when the true mean difference is zero*, the magnitude of an observed mean difference will exceed 1.96 times (say, twice) its standard error with probability less than 0.05 and will exceed 2.58 times its standard error with probability less than 0.01.

An Aside

We are using the normal in place of a related distribution (called Student's t-distribution). The correct multipliers with which to replace 1.96 and 2.58 can be derived from this distribution but vary with sample size. Table 5.2 shows the extent of our approximation as n varies. The above multipliers always *undervalue* the true tail probability.

Probability distributions such as these are central to statistical thinking and the intensive use of mathematics as a tool in their derivation has led many to the conclusion that statistics is merely a branch of mathematics. Had electronic computers been around 200 years ago, it is likely that statistics would have become more closely associated with computing science, for the computer provides a more powerful means of determining to any required accuracy the shape of a probability distribution by simulating the mechanism of interest. Thus if the mathematical formulation of the binomial distribution were not known to

Table 5.2 Probability with which a sample mean differs from the population mean by at least 1.96 and 2.58 standard errors for increasing sample size.

Sample size n	Prob. < 1.96 × st. error (%)	Prob. < 2.58 × st. error (%)
10	8.57	3.26
20	6.57	1.89
30	6.00	1.54
60	5.48	1.24
100	5.28	1.14
300	5.09	1.04
infinite	5.00	1.00

us, we could set a computer to generate millions of 'throws' of a die and so build up a more or less correct picture of the probability pattern. This technique of **simulation** has assumed considerable importance in modern statistics in the solution of problems that are mathematically intractable.

All that need concern us here is that, whether the probability pattern is derived mathematically or established by simulation, the end result is the ability to predict the relative frequencies of observable values of sample statistics resulting from a known state of nature.

5.4 TESTING A HYPOTHESIS

As we have already hinted, the need for the idea of the probability distribution arises from the sort of back-to-front argument that is often used in statistics to make an inference about a population on the basis of a random sample of its members. The approach, known as **hypothesis testing**, is akin to what logicians call *reductio ad absurdum* or 'indirect proof', exemplified by the following argument:

(a) 'Let us hypothesize that at the time of the accident the car's braking system was totally inoperative.'
(b) 'We therefore deduce that no skid marks would have been found.'
(c) 'But skid marks were found.'
(d) 'Hence our original hypothesis is wrong.'

The logic of this argument is inescapable (though you may feel inclined to say 'Yes, but...', in which case you should make a good statistician!). And notice most importantly that, if skid marks had not been found (step (c)), we could have said nothing for or against our hypothesis.

Now let us look at the statistical version of these four steps as applied to an investigation of whether a particular die is biased in favour of four:

(a) We begin by hypothesizing that the die is not biased; this is usually called the **null hypothesis** since it suggests a neutral state of nature – very often the reverse of what we hope to prove. We are going to ask someone to throw the die 30 times and we assume this individual has no control over the fall of the die. This assumption has nothing to do with the null hypothesis but is none the less important.

(b) The logic of Section 5.3 and, in particular, the table of the binomial distribution for 30 throws of a true die now tell us the probabilities for the number of fours that might appear *if the null hypothesis is true.*

(c) Now we do the experiment. If the number of fours turns out to be much higher than would have been expected under the null hypothesis (i.e. 5) then our suspicions of bias in the die will be aroused. Suppose ten fours appear. The degree of our suspicion will be measured by the probability of finding an observed value of the sample statistic at least as unfavourable to the null hypothesis i.e. from Table 5.1 with $n = 30$ and $p = 1/6$,

$$\begin{aligned}\text{Prob}(10) + \text{Prob}(11) + \text{Prob}(12) + \cdots + \text{Prob}(30) \\ = 0.0131 + 0.0048 + \cdots + 0.0000 \\ = 0.0199\end{aligned}$$

i.e. if the null hypothesis is true, we would see ten or more fours in just under 2% of such experiments.

(d) Hence one (or more) of three things must be true:
   (i) an unusual event has indeed occurred
   (ii) our assumption about the thrower is false
   (iii) the null hypothesis is false.

We cannot know for sure which of these to believe – that is part of the uncertainty of natural variation. However a number of comments can be made:

1. We can at least think very hard about the assumption, maybe even find a way of checking it. Assumptions in the testing of hypotheses are sometimes related to necessary statistical conditions such as the randomness of a sample, sometimes to the commonsense requirements of good study design. These assumptions should always be listed prior to data collection and, if possible, checking in some way or at least discussed as acceptable and unlikely to invalidate the main hypothesis if wrong. Many fallacious statistical arguments arise, not through incorrect use of methods, but through overlooking some vital assumption.
2. Conventionally, a probability determined in step (c) above as less than 5% (1% in medical research) is regarded as providing sufficient evidence to *reject the null hypothesis.* It might be said, for example, that our little experiment has provided *considerable evidence* that the die is biased. However, we certainly have not proved that it *is* biased; our findings cannot be as clear cut as with the

argument about the skidding car. Nevertheless, the lower the probability of the experimental result, the stronger is the evidence against the null hypothesis. The technique under discussion is sometimes called **significance testing**; it would be said that we have rejected the null hypothesis of an unbiased die at the 2% level of significance because that is the percentage point below which our calculated probability has fallen. The word 'significant' has therefore a more specialized meaning for a statistician than for the lay person.

3. If the experimental result had not been particularly unusual for an unbiased die, indeed, even if exactly five fours had been observed, we could not have accepted that the die *is* unbiased. Such an incorrect term is sometimes used but it would be as illogical to accept that the car's brakes were inoperative if no skid marks had been found. The need for action may sometimes lead us to proceed as if a null hypothesis were true even though the logical test of the hypothesis is inconclusive.

Looking at the fairness of a die is instructive but hardly demonstrates the utility of hypothesis testing in a real-life context. We now consider examples of five less frivolous applications.

Example 5.1

In an attempt to find people with an ability to judge cheeses, a subject is given a so-called 'triangle test'. In each of 30 trials, he is asked to taste three cheeses – two identical and one slightly different – and has to indicate the odd one out. If he is good at discriminating, he will be correct more often than in the $30 \times 1/3 = 10$ trials that random choice would suggest. Suppose he is correct 16 times. The logic of our significance test of his ability would proceed thus:

(a) Our null hypothesis is that he has no ability to discriminate among cheeses. We assume also that the experiment is well conducted (for example, that there is a reasonable time lapse between the trials and the subject washes out his mouth after each tasting), that the trials are conducted under conditions typical of those in which his possible skill might be used (time of day, room temperature, etc.), that the subject tells the truth, and so on.

(b) Under the null hypothesis, the number of correct decisions will be an observation from the binomial distribution with $n = 30$ and $p = 1/3$.

(c) The probability of his being correct at least 16 times if he truly has no tasting ability can therefore be calculated from Table 5.1 with $n = 30$ and $p = 1/3$ as

$$\text{Prob}(16) + \text{Prob}(17) + \cdots + \text{Prob}(30)$$
$$= 0.0115 + 0.0047 + \cdots + 0.0000 = 0.0185$$

(d) As with the die, a rather low probability has been found. If we can discount any flaw in our assumptions and do not accept that the subject's high score could be due to chance (maybe a mistaken belief) then we must conclude that our null hypothesis is false and that the evidence suggests that the subject does have some tasting ability.

Example 5.2

Two types of weedkiller, A and B, were compared in the control of bracken using 10 pairs of adjacent plots rated by an expert as being equally infested with the weed. The herbicides were applied, one to each plot of a pair, and the following year the expert judged that the A plot had less remaining bracken than the B plot in 8 of the 10 pairs. Is this evidence of a difference in efficacy of the two herbicides or could the result be due to chance?

(a) Our null hypothesis is that there is no difference in the effectiveness of A and B. We assume also that there were no external influences that favoured one of the treatments, that the expert's opinion was unbiased, and so on.

(b) Under the null hypothesis, herbicide A would produce a greater reduction than B about 50% of the time. The number of plots where A is superior would therefore be an observation from the binomial distribution with $n = 10$ and $p = 1/2$.

(c) The probability of A being better than B in at least three more plots than one might have anticipated under the null hypothesis is, from Table 5.1 with $n = 10$, $p = 1/2$,

$$\text{Prob (8)} + \text{Prob (9)} + \text{Prob (10)}$$
$$= 0.0439 + 0.0098 + 0.0010 = 0.0547$$

To this must be added the probability of A being worse than B in at most two plots since such a result would have been just as much at odds with the null hypothesis. This probability is, by the symmetry of the distribution with $p = 1/2$, also equal to 0.0547 so that, overall, we can say that, if the null hypothesis were true, a discrepancy between the performances of the two herbicides at least as great as has been found would occur in about 11% of such experiments.

(d) Such a probability is so high as to be regarded as reasonably consistent with the null hypothesis; we can say that the difference in effectiveness is not significant. Of course, we have not proved that the two weedkillers *are* equally useful; we have merely failed to find evidence of a real difference (though such a difference may exist and could have been detected by a larger experiment).

Example 5.3

A new liver transplant technique has been developed and, of the first seven patients treated, two are still alive three years after their operations, the remainder having survived for

32 13 28 19 4

months. Are these figures consistent with a *median* survival of 6 months? (Recall Section 3.1.2.) We look at the median rather than the mean because the survival distribution is likely to be skew.

(a) The null hypothesis is that, in the notional population of all possible liver

transplant patients, half will survive for at least six months. Many assumptions are necessary about this population, about the seven patients in the study and about the treatment itself; we leave the reader to think about these.

(b) Under the null hypothesis and assumptions the number of patients who survive six months or longer is an observation from the binomial distribution with $n = 7$ and $p = 1/2$.

(c) A total of six patients survive for six months or more; the probability of at least six out of seven exceeding the median survival time is, from Table 5.1 with $n = 7$, $p = 1/2$

$$\begin{aligned} \text{Prob }(6) + \text{Prob }(7) &= 0.0547 + 0.0078 \\ &= 0.0625 \end{aligned}$$

We must also recognize that the identical (because, in this case, $p = 1/2$) probability of finding six or seven patients with survivals lower than the median would equally provide evidence against the null hypothesis. If the null hypothesis were true, therefore, a pattern of survivals such as we have found has an associated probability of 0.1250.

(d) This relatively high probability gives no evidence against the hypothetical median survival of six months. Notice that the data are equally consistent with the true median survival lying anywhere above 4 and less than 13 months.

Example 5.4

Recall that in Section 3.6 we stated that an observed difference of 0.67 miles per gallon in the mean petrol consumption of two makes of car was 'not spectacular' when related to a standard error of 0.480. This remark can now be explained in hypothesis-testing terms.

(a) We hypothesize no difference in mean fuel consumption for the two makes of car. We assume that the cars used were random samples from the two production lines and that the conditions of data collection were fair and sensible.

(b) The number of experimental units is arguably sufficiently large that we can say that an observed mean difference will have an approximately normal distribution.

(c) The observed difference of 0.67 is a good deal less than twice its standard error of 0.480. The probability of finding such a difference if the null hypothesis were true is therefore well over 5%. (Actually, it's 16.3%)

(d) Provided the assumptions are acceptable, we can therefore say that the difference in performance is 'not spectacular', i.e. we have no evidence to cast doubt on the null hypothesis. We emphasize again, however, that a sufficiently large experiment could well detect a difference; we certainly cannot say that the mean fuel consumptions of the two makes of car *are* identical.

Example 5.5
Thirty subjects were given a problem-solving test, their completion times being measured before and after receiving a standard dose of a drug (? alcohol); this is therefore a paired comparison study (see Section 3.6.1.). The results are given in Table 5.3.
Was average performance affected by the drug?

(a) The null hypothesis is that of no difference in average completion times. We assume that the subjects were a random sample of the population to which the findings are to apply, that they had had an earlier phase of familiarization with the test procedure and that the circumstances of the two timings were the same.
(b) The sample size is sufficiently large for the observed mean difference to have an approximately normal distribution.
(c) The reader can confirm that the mean of the (after − before) mean difference is 0.490 which is therefore just less than 2.58 times its standard error of 0.1911. The probability of such a discrepancy if the null hypothesis were true is thus very little over 1%.
(d) Thus we have considerable evidence of a difference in mean times.

Table 5.3 Effects of drug ingestion on completion of a problem-solving task

	Times (secs)				Times (secs)		
Subject	Before	After	Diff.	Subject	Before	After	Diff.
1	9.60	12.02	2.42	16	8.32	9.13	0.81
2	10.82	11.36	0.54	17	11.58	10.35	−1.23
3	5.47	6.20	0.73	18	8.75	7.89	−0.86
4	7.94	7.64	−0.30	19	9.53	10.10	0.57
5	7.82	8.84	1.02	20	9.78	9.45	−0.33
6	13.42	14.54	1.12	21	11.83	12.29	0.46
7	7.87	10.35	2.48	22	10.75	12.73	1.98
8	8.71	9.61	0.90	23	9.22	8.44	−0.78
9	10.58	11.85	1.27	24	7.74	6.74	−1.00
10	11.93	11.95	0.02	25	7.69	8.71	1.02
11	10.41	11.92	1.51	26	10.32	8.88	−1.44
12	11.89	13.15	1.26	27	8.94	9.55	0.61
13	8.02	9.33	1.31	28	5.01	4.55	−0.46
14	9.89	11.37	1.48	29	10.51	11.07	0.56
15	9.45	9.15	−0.39	30	12.04	11.47	−0.57

An aside

If we believe that the differences come from a symmetric distribution, we could equivalently test for the median difference being zero. Under that null hypothesis, the number of negative differences (i.e. 10) is an observation from the binomial distribution for $n = 30$ and $p = 1/2$ with associated probability (from Table 5.1) of

$$0.0280 + 0.0133 + 0.0055 + 0.0019 + 0.0006 + 0.0001 = 0.0494$$

This must be doubled because 10 *positive* differences would be equally discrepant, i.e. we could have chosen to take the (before − after) differences. Hence, the overall probability is 9.88% – no evidence of a difference in median completion times. This test – called the **sign test** – appears to give a different conclusion from the previous one. However, this should not be too surprising; since we have thrown away the information contained in the *magnitudes* of the differences, the test must clearly be less powerful. The sign test is typical of a class of tests that are (not well) described as **non-parametric** because the magnitudes of observations are ignored; they are generally less powerful but require fewer assumptions than equivalent parametric tests. Notice that Example 5.2 was really an application of the sign tes

---

5.4.1 Alternative hypothesis

In Example 5.1, we computed the probability associated with only one tail of the sampling distribution; for all the other examples, both tails were calculated. The reasoning behind this derives from the idea of the **alternative hypothesis**. In the triangle test, p (the probability of a correct judgement) would equal 1/3 if the subject has no skill or would exceed 1/3 if he or she does have some skill. It does not make sense to consider $p < 1/3$ as an alternative; low observed values of success rate cannot therefore be regarded as evidence against the null hypothesis and the test is said to have a **one-sided** alternative. This situation is somewhat unusual. In most cases (including the other four examples), the direction of possible alternatives to the null hypothesis is non-specific and extreme values of the test statistic in either tail must be included in evaluating the contrary evidence – a **two-sided** alternative.

5.5 PROBLEMS OF HYPOTHESIS TESTING

We have considered in dètail five different examples of the use of hypothesis testing in simple cases. A glimmering of where this could lead may impress some readers. Others, on flicking through the pages of most statistics textbooks or glancing at the average statistics examination question, may be alarmed at the pervasive nature of the idea. Those who have occasion to read learned journals in science and medicine will be amazed at the way editors of such publications have come to insist that authors incorporate a sprinkling of significance tests in the belief that scientific precision is thereby demonstrated, even where assumptions are never discussed nor the appropriateness of the tests considered.

It is hoped that the material in this book will help readers to think constructively about the assumptions that are inherent in most scientific research. Hypothesis testing has achieved its prominence in statistical analysis partly because many people are unhappy if they cannot conclude with a yes/no answer, partly because such a relatively easy-to-apply recipe is seen as a welcome substitute for some original thought, and partly because examination questions in the topic are easy to produce. But there are a number of good reasons why the practical person, having tested a hypothesis, should stand back and consider the relevance of what he or she has done.

In the first place, a comment about the nominal significance levels of 5% and 1% is in order. The arbitrary nature of these cut-off points should be evident and it would clearly be ridiculous either to ignore a probability level of 5.1% or to cry 'Eureka!' over a probability of 4.9%. The actual probability level should be reported whenever possible, but this is particularly so in the case of discrete sample statistics of which the binomial count is an example. Even with n as large as 30 (and $p = 1/2$), the reader can calculate that the jump in going from counts of 9 or 21 to 10 or 20 is from 4.28% to 9.88%. The gap caused by a change of 1 in the observed count (and it could be much greater for smaller n) should underline the dangers of nominal significance levels.

Secondly, most investigations have the potential to answer several questions rather than simply a single one. For example, an industrial experiment on the production of petrochemical products may well compare dozens of combinations of temperature, time, amount of catalyst, and so on, as to their effect on yield of first- and second-stage derivatives, amounts of various effluents, noxious gases, etc.; the number of possible tests is very considerable and they are all interrelated so that the probability of incorrect rejection of a null hypothesis is much greater overall than one might suppose. The same sort of difficulty will beset a clinical chemist looking for the existence of associations among the 120 possible pairs of the 16 blood chemicals commonly determined; he will find too many exciting but unreal relationships. The problem has become exacerbated by the arrival of the electronic computer which, with unintelligent (so far) but assiduous zeal, automatically bombards the user with every possible significance test, relevant or not. It must be intuitively obvious to the reader that this is a sure-fire recipe for finding fool's gold. There are means of resolving some of the multiple tests problems but they are not always satisfactory even in the hands of the professional statistician.

The next remark to be made is that the evidence from a single experiment or survey should never be considered in isolation. For instance, EEC rules on the acceptance of new crop varieties demand that trials be made at several widely differing sites over several seasons. Drug companies likewise attempt to substantiate preliminary findings by collaborative multi-centre trials using a wide variety of patients. Although the expert can often combine evidence from different studies in a meaningful way, consistency in the findings is generally more vital to wide acceptance of claims made than any isolated test of a null hypothesis.

Lastly, and perhaps most importantly, the point at issue is not usually whether or not such-and-such an effect exists, but rather what is its magnitude. No one believes, for instance, that two genetically different varieties of potato can have identical average yields and, indeed, a sufficiently large scale comparison would be able to reject a null hypothesis of equality; but to persuade a farmer to change to the newer and doubtless more expensive variety one would need to be able to quantify his expected improvement in yield. On the other hand, a migraine sufferer, told that a new treatment was better than the old in 19 out of 30 cases (which, as the reader can now confirm, provides little evidence of its real superiority), would nevertheless be tempted to switch regimes, other things being equal. What we are saying, in summary, is that statistical significance is largely unrelated to practical importance.

5.6 ESTIMATION

From what has been suggested in the closing paragraph of the last section, the reader will appreciate the need for some more positive means of gauging the true state of nature than the 'Aunt Sally' approach of significance testing. This need is largely met by the process of **estimation**, the full details of which are slightly outside the scope of this book. Nevertheless, it will be helpful to introduce the basic ideas.

It is intuitively clear that a sample statistic such as a mean or proportion must, in general, provide some indication of the corresponding population figure and that, the larger the sample, the greater is the confidence in the value inferred. Such a sample statistic is called an **estimate** of the population analogue. It is in this sense that one can regard 1.2 decibels as an estimate of the true difference in noise level of two types of aircraft whether or not the samples that gave rise to this figure indicate a significant disparity.

One would hope that, over all possible samples, the average value of the estimator would be the true population value and, for most elementary estimators, this property can be demonstrated. One could then reasonably ask that, for a given sample size, the estimator should have as small a variance as possible. As we have already remarked in Section 3.2, this favours the mean over the median as the esimator of the centre of a symmetric distribution.

But, however small the variability of an estimator, it is unlikely that the estimate obtained in any particular sample will actually equal the true population figure. It will therefore generally be more useful to determine a range that is likely to include the true value. Such a range is called a **confidence interval**. The end points of this interval will be random variables with values derived from the observed sample in such a way that the probability that they bracket the value to be estimated is, say, 95% or 99%. For means or mean differences *based on large samples*, for instance, the lower end point for the 95% confidence interval is given by

$$\text{observed mean (difference)} - 1.96 \times \text{standard error}$$

and the upper end point by

$$\text{observed mean (difference)} + 1.96 \times \text{standard error}$$

with 1.96 being replaced by 2.58 for the 99% interval.

Thus, if an observed mean noise level difference of 1.2 decibels based on large random samples of take-offs of two types of aircraft has a standard error of 1.33 decibels, the 95% limits would be

$$1.2 - 1.96 \times 1.33 = -1.41$$

and

$$1.2 + 1.96 \times 1.33 = 3.81$$

and we could therefore claim to be 95% confident that the range (−1.41, 3.81) decibels includes the true mean difference in noise level during take-off for the two types of aircraft. Incidentally, the fact that this range includes zero is sufficient to tell us that there is no significant difference (at the 5% level) in mean noise so that, even from this brief introduction, the reader will discern that a confidence interval carries more useful information than the corresponding hypothesis test. The reader may care to check that the 99% confidence interval is (naturally) wider at (−2.23, 4.63) decibels. We emphasize again, however, that these confidence intervals are valid *only for large samples.*

SUMMARY

The variability of sample statistics over different samples indicates a major difficulty in inferring the true nature of a population. One response to the problem is to measure our surprise at the observed value of a statistic if a particular hypothesis about the population is true. The degree of surprise is quantified in terms of the relative probability with which the different possible values of that sample statistic would appear. A more useful approach is provided by estimation and the construction of a confidence interval. In both cases, assumptions are usually essential and this further component of uncertainty must be acknowledged and honestly considered.

EXERCISES

1. Randomly selected 'witnesses' (292 males and 296 females) were each asked to judge the height (inches), weight (1b) and age (years) of a 'target' person they had seen a few seconds earlier and of a 'context' person whom they had seen with the target and who asked for the witnesses' judgements. (Target/context pairs were selected haphazardly from 14 experimenters.) Differences computed as (true value − estimated value) were summarized as shown in Table E5.1.

   (a) Test for biases in the witnesses' judgements.
   (b) What further information would you need to test whether context judgements are as biased as target judgements.
   (c) Test whether biases are the same for male and female witnesses.

Table E5.1 Judgement of height, weight and age

	Male witnesses ($n = 292$)		Female witnesses ($n = 296$)	
	Mean	St. devn.	Mean	St. devn.
Target				
Height diff.	2.538	2.525	3.348	2.549
Weight diff.	3.904	19.348	8.957	21.374
Age diff.	−2.541	3.413	−2.642	3.272
Context				
Height diff.	2.229	2.039	3.331	2.664
Weight diff.	5.449	19.251	11.584	19.527
Age diff.	−1.842	2.858	−1.983	3.071

2. (a) A research publication reports that a large random sample of normal males produced a 95% confidence interval for mean serum cholesterol of (214.12, 225.88) mg/100 ml. For a large sample of men with a genetic abnormality the corresponding interval was (225.66, 241.34) mg/100 ml. These confidence intervals overlap but this does not indicate that the two means are not significantly different; show that, on the contrary, they differ at the 1% level of significance.
   (b) Another medical report on blood pressure measurements (in mm Hg) gave the 95% systolic pressure confidence interval based on a large sample of people measured at home as (145.2, 157.0), while the same sample measured at a clinic gave (152.8, 168.4). Why is it impossible to test the difference in mean systolic pressure for the two environments using these figures ?

3. [Preliminary
   The standard error of a sample regression coefficient is

$$\mathrm{SE}(\hat{\beta}) = \sqrt{\frac{\mathrm{RMS}}{S_{xx}}}$$

The standard error of the difference between a predicted $\hat{y}_0$ at x_0 and a new value, y_0, observed at x_0 is

$$\mathrm{SE}(\hat{y}_0 - y_0) = \sqrt{\mathrm{RMS}\left(1 + \frac{1}{n} + \frac{(x_0 - \bar{x})^2}{S_{xx}}\right)}$$

For large n, both $\hat{\beta}$ and $\hat{y}_0 - y_0$ have approximately normal distributions; if their true (population) values are zero, their magnitudes will exceed 1.96 or 2.58 times their standard errors with probabilities 0.05 and 0.01 respectively. This allows us to test whether β could be zero (equivalently, whether the correlation between x and y is zero) and whether the new (x_0, y_0) pair is consistent with the previous data.

Refer back to Exercise 1 of Chapter 4 and Exercise 4 of Chapter 2.

(a) Is SCK level significantly correlated with age in
  (i) the normal women,
  (ii) the Becker carriers?

(b) A woman aged 25 is found to have an SCK level of 56. Could she be a carrier? Could she be normal?

4. (*See preliminary to Exercise 3.*)
For a random sample of 100 statistics students, the regression of statistics mark (y) on IQ ($x; \bar{x} = 116.2$, $S_{xx} = 11\,449.4$) gave

$$\text{stats\%} = 30.5 + 0.244 \times \text{IQ}$$

with residual mean square = 80.885

(a) Are IQ and statistics performance linearly related? What is the correlation between them?

(b) Of 20 students examined the following year, only one with an IQ of 130 and statistics mark of 85 appeared to be significantly out of line with the predicted result. Maybe we have evidence of cheating; maybe the digits of the examination result were accidentally transposed; maybe the years are not comparable for some reason; maybe nothing untoward has happened. Discuss.

5. Binomial tail probabilities are somewhat tedious to calculate for large values of n. Luckily, the normal distribution comes to the rescue by providing an approximation that is valid as long as n is reasonably large and p is not too near to 0 or 1. (Indeed, historically, this was the first use of the normal distribution.) Let

$$z = \frac{f - p - g/2}{\sqrt{gp(1-p)}} \qquad \text{where } f = \frac{r}{n} \text{ and } g = \frac{1}{n}$$

The following table gives the probability of finding a normally distributed observation equal to or less than z when $n = 30$ and $p = 1/2$.

r	prob $\leqslant z$	r	prob $\leqslant z$	r	prob $\leqslant z$
5	0.0003	12	0.1807	19	0.9498
6	0.0010	13	0.2919	20	0.9777
7	0.0031	14	0.4276	21	0.9912
8	0.0088	15	0.5724	22	0.9969
9	0.0233	16	0.7081	23	0.9990
10	0.0502	17	0.8193	24	0.9997
11	0.1006	18	0.8994	25	0.9999

(a) Compare these tail probabilities with the probabilities of an observed count being equal to or less than r in the corresponding binomial distribution.

(b) To use this normal approximation to test whether a population proportion could have a specific value, p_0, given an observed proportion of $f(= r/n)$, the *magnitude* of $(f - p_0)$ is reduced by $g/2$ and compared with its standard error of $\sqrt{gp_0(1-p_0)}$ where $g = 1/n$. In an obscure die-tossing experiment in 1894, W.F.R. Weldon asked a clerk to record the number of scores of 5 or 6 in 84 072 throws. The clerk reported 28 307 such results; should Weldon have been suspicious about his honesty?

(c) If samples of sizes n_1 and n_2 from two populations give observed proportions of f_1 and f_2 in a certain category, is this evidence that the population proportions differ? Let f be the proportion in the category over both samples, i.e.

$$f = \frac{n_1 f_1 + n_2 f_2}{n_1 + n_2} \qquad \text{and let } g = \frac{1}{n_1} + \frac{1}{n_2}.$$

Then, provided n_1 and n_2 are each at least 30 and $n_1 f$, $n_2 f$, $n_1(1-f)$ and $n_2(1-f)$ are all at least 5, the *magnitude of* $(f_1 - f_2)$ reduced by $g/2$ has approximately a normal distribution with standard error $\sqrt{gf(1-f)}$. Other data produced by Weldon gave 78 295 scores of 5 or 6 out of 231 600 throws. Are the two sets of data consistent?

6
Published statistics

6.1 THE COLLECTION OF DATA

In the conventional view, data are collected by the sponsor of a study to attempt to answer the specific questions posed.

This may require that an **experiment** be undertaken where the variables of interest are to a greater or lesser degree under the control of the researcher. Usually a number of variables are set at predefined levels and the resulting values of one or more 'response' variables observed; this would be the situation in an industrial experiment to study the variation in yield of some product resulting from changing levels of temperature, amount of catalyst, process time, and so on. Such an approach allows us to experience nature's response to the controlled variation of an environment. Indeed, the French call an experiment 'une expérience'. We consider experimental method further in Chapters 15 and 16.

Alternatively, a **survey** may be used to gather data in circumstances where little or no control over the value of variables of interest is possible. This is usually the case for studies of human beings whose lifestyles cannot generally be mainipulated except possibly in some short psychological experiment or to further medical research. But a survey is often the only approach even for the physical sciences; an engineer, investigating the effects of gusts of wind on a suspension bridge must accept the wind speeds that nature happens to provide during the period of observation.

A third means of finding data for certain areas of research, particularly but not exclusively in the social sciences, is to look at information previously collected by someone else for their own purposes. In this country, the Economic and Social Research Council frequently awards research grants only on condition that data collected will be deposited in the Council's archive together with a full description of the material and any known deficiencies. A good deal of government-collected data are also held. Such archived information is frequently used in research because it is more or less free; but, to avoid fallacious conclusions arising from what is usually called **secondary analysis**, we must be prepared to think critically about the source, reliability and consistency of the data involved. In particular we need to check that the population defined and the sample extracted by the originator of the material are consonant with the objectives of our own study and that there is no ambiguity in any technical terms or schemes of coding. Sadly, even information specifically stored for secondary analysis can be marred by obscurities in definition that are not easily resolved.

A somewhat different aspect of the re-use of data involves **published summary tables**. As time goes by, more and more information is being routinely gathered on more and more aspects of life by government departments and international agencies. This 'growth industry' feeds on accelerating social change and the increase of centralized planning and control and is aided by the computer's ability to store vast amounts of data and produce extensive tabular summaries. Because figures published by government bodies are usually of high quality, we shall concentrate on these in what follows although other data sources are mentioned. We begin with a brief and necessarily incomplete review of UK statistics; the general provisions are typical of what is available in countries of the Western world.

6.2 UK PUBLISHED STATISTICS

Readers are reminded that the term 'United Kingdom' (UK) includes Northern Ireland, whereas 'Great Britain' (GB) does not. Of the constituent countries of the UK, Northern Ireland had a considerable degree of devolved government but this has been modified in recent years pending resolution of Ireland's partition problems. Scotland has no separate parliament but differs from England so greatly in law and education that a fairly autonomous civil service operates from Edinburgh (the Scottish Office) and there is also a separate Registrar General. Wales is almost always treated as though it were an offshoot of England although there is a small group of civil servants in Cardiff for specifically Welsh matters. Needless to say, all this is very confusing to outsiders!

6.2.1 The census (head counting)

The 1920 Census Act specified that population enumerations should take place at intervals of not less than five years, conducted by the Registrars General of England and Wales, of Scotland and of Northern Ireland. In fact there has been a complete census in the first year of every decade since 1801 (excepting 1941). Owing to faster population growth and an internal migration rate approaching 10%, an additional 1 in 10 sample was taken in 1966 and this would have been repeated in 1976 but for economic stringencies.

The Act lists basic questions to be asked but permits census-to-census variation and often a census will concentrate on some topic of current interest. For instance, the censuses of 1971 and 1981 investigated country of origin more fully than had been done previously to extend knowledge about the immigrant population.

In many countries, notably the USA, censuses are *de jure*, i.e. a person is enumerated in respect of his or her 'usual' residence. Although in 1981 a question was asked about usual residence, UK censuses are *de facto*, i.e. individuals are enumerated on the basis of where they are at the time of the census. (This avoids the need to define the distinction between temporary and permanent residence). The method is least satisfactory in health resorts, coastal ports, London etc., and it is obviously necessary to avoid peak holiday times and advisable to choose a weekend. A Sunday in April is the usual compromise. It has been estimated that *de facto* registration inflates the total by about 0.2% because of foreign visitors and that about 2% of the true population are displaced.

In Great Britain there are around 2300 census districts each in the charge of a census officer (usually a member of the local registration staff) who controls specially recruited enumerators – one for each of the 110 000 or so enumeration districts. All are given extensive training to clarify their duties. The enumerator must deliver a schedule to the head of each household (including caravans, tents, etc.) in his or her district a few days before census day. This should be completed by the head of the household *on* census day and it is collected by the enumerator within two days, any queries being sorted out. The forms are checked and returned to the census officer for coding. Enumerators summarize some basic figures and these can be used to prepare preliminary tables for early publication. Finally external checks on accuracy (e.g. by small-scale postal survey) are made.

6.2.2 Vital registration (event counting)

The registration of births, marriages and deaths has been compulsory since the middle of last century and is used to supplement census figures in intervening years.

Births must be registered within 42 days; the information includes date and place of birth, name (if decided), sex, name of father, maiden name of mother, father's occupation. (In Scotland, additional information on date and place of marriage allows easy construction of family trees which is vital in genetic research.) As an encouragement to speedy registration, payment of child allowance is made dependent on production of the birth certificate. Since 1926 (1939 in Scotland) stillbirths (over 28 weeks gestation) have been compulsorily recorded together with cause.

Death being a fairly serious matter, a body cannot be disposed of until a doctor's certificate as to cause of death has been received by the registrar. This should be within 5 days of death. The information recorded is date and place of death, name, sex, age, occupation (of husband for a married woman), cause of death, marital status, age of surviving spouse if married.

Marriage has both religious and legal aspects and this results in a complicated registration system. The information recorded is date, precise place of marriage and form of ceremony and, for both parties, name, age, marital status, occupation, residence and father's name and occupation.

6.2.3 Special sample surveys

Some examples are given below

(a) Family Expenditure Survey

This involves a survey of about 11 000 households per year from all over the UK to obtain information on personal income and expenditure. The interviews take place throughout the year to take account of seasonal variation. The information collected is used in the construction of the retail price index (see Chapter 8).

(b) General Household Survey (since 1971)

About 13 000 households are interviewed each year in Great Britain, the questions covering constantly varying subjects such as family composition, housing, unemployment, illness, and medical attention, long-distance journeys, educational qualifications, etc. Once again, the interviews take place throughout the year.

(c) International Passenger Survey

A survey is made of a random sample of about 100 000 passengers into and 125 000 passengers out of main air and sea ports. (A migrant is one intending to stay out or in for more than 12 months.)

6.2.4 Continuous Population Register

Such a register was available in the UK only from 1939 to 1952, but they have been widely used in, for example, Scandinavia (since the seventeenth century), Belgium, the Netherlands and Switzerland. A record (nowadays computerized) is generated at birth and information on marriage, children, internal migration, change of occupation, etc., is added throughout the life of the individual. This is clearly very useful but administratively difficult and expensive. All the countries mentioned above continue to hold censuses to check against deterioration in quality of data held.

6.2.5 Health statistics

Partly because of a worldwide interest in comparative health studies, statistics on health are a major component of UK government publications.

(a) Hospital Inpatient Survey

This is a 1 in 10 sample in England and Wales – a complete census in Scotland – at discharge or death. It is used for administrative purposes (bed usage, duration of stay, operations, etc.) and studies of illness treated in hospital.

(b) Special databases

Several databases have been established to provide continuous monitoring of the health of individuals, e.g. the Nottingham Psychiatric Case Register, the Medical Research Council General Practice Patient Register (in Edinburgh, for genetic studies). These are not government controlled.

(Occasional controversy in the press has caused public concern that medical records held by computers are less confidential than those left lying around general practice receptionists' rooms and hospital records offices. For example, fear exists that growing computerization of personal records might enable a potential employer (particularly the Health Service) to have access to private medical information and, for this kind of reason, the National Insurance Number, which might otherwise be a useful patient identifier, is never used as such. As a result of these fears, the Data Protection Act which came into operation in 1986 seeks to regulate the ways in which personal data can be stored by electronic means and gives the individuals concerned rights to challenge accuracy.)

(c) General Household Survey

(See Section 6.2.3(b).)

(d) Other special surveys

There are many regular or occasional studies of, for example, psychiatric illness, casualty services, handicapped children, cancer incidence, and so on, sponsored by research councils, health boards, academic institutions and drug companies.

6.2.6 Education

An enormous amount of information is now collected on the performance of individual pupils throughout their school and further education careers and on numbers and grades of teaching staff, e.g. the Universities Statistical Record held on the UCCA computer. (Elaborate confidentiality precautions are taken.)

6.2.7 Crime and justice

Extensive amounts of data are collected on convictions, the prison population, children in care, etc.

6.2.8 Publications

The publications resulting from these collected data are far too numerous to list individually though such a directory is available. The main sources are:

1. Annual Reports of the Registrars General for England and Wales, Scotland and Northern Ireland.
2. *Annual Abstract of Statistics* – published by the Central Statistical Office and covering not only current figures but the preceding 10 years.
3. *Monthly Digest of Statistics* – mostly economic and financial figures.
4. Special publications for Scotland, Wales and Northern Ireland, e.g. Highlands and Islands Development Board reports.

6.3 RELIABILITY

Often common sense or investigation of the method of data collection will reveal that not all data are equally 'hard' (i.e. reliable). There can be no boundary between soft and hard information – being aware of the problem in relation to the proposed use of the data is what matters. Generally, government publications (in this country) are reliable, though even civil service reports are censored against the presence of politically sensitive material, e.g. number of coloured people. Because constituency boundaries are fixed on the basis of figures produced by the Registrars General, these officers are not subject to ministerial control and their publications are more likely to be free of such problems. Indeed the census total usually turns out to be very close to the estimate derived from the previous census figure corrected for intervening births, deaths and net migration. Nevertheless, it will be appreciated that, for example, in the census, the main obligation for accuracy rests on the head of the household, who may intentionally or unintentionally provide false information. Random errors will not matter greatly; what is of concern is the possibility of bias. For instance, there is a tendency for those with a birthday imminent to return the higher age instead of the attained age, whereas middle-aged women still tend to understate their ages; this phenomenon is sometimes called 'age-shifting'. There are inordinately large numbers of ages ending in 0 and 8 and, to a lesser extent, 5; this is known as 'age-heaping'. Memory errors resulting from questions such as 'Did you live in your present house five years ago?' are particularly troublesome.

Occupations tend to be upgraded in status, for example, by implying possession of a special skill or suggesting that the work is in a supervisory capacity. Those who are unemployed or retired tend misleadingly to give their former occupation.

Of the information collected in vital registration, cause of death is obviously least reliable. Doctors will often differ as to cause, particularly since 'died of old age' is not a permitted entry; the unreliability is therefore greatest for the oldest age groups. Furthermore, since up to three causes are allowed for in the record, the doctor may indicate, say, pneumonia as 'primary' (i.e. immediate) cause before a more basic diagnosis of lung cancer. Indeed fashions in specifying cause of death change from time to time so that the entry may even depend on when the doctor was trained!

A general problem will be apparent in studies of, for example, hospital in-patient statistics where repeated admissions of the same individual are multiply counted. Thus, an illness that tends to require frequent short spells in hospital will appear to have a high incidence. In theory, such admissions could be linked by surname, date of birth, address, etc., but, in practice, errors in recording these items make this very difficult. In any case, hospital admission depends locally on waiting lists and varying medical policy. Furthermore, since the record is generated on discharge, complications relating to length of stay will disturb calculated sickness rates.

Criminal statistics are notoriously unreliable. For one thing, we can only guess at the proportion of actual offences known to the police. This varies with the nature of the offence (murder versus parking on a yellow line), the strength and efficiency of the police and the recording methods employed (under-reporting of offences by young children, pregnant women and old people). Even with detection and reporting, a prosecution may not follow; failure to prosecute is more marked in Scotland where requirements of corroboration are higher than in England and Wales. However, this is almost more a problem of consistency.

6.4 CONSISTENCY

It is vital when *comparing* two sets of published figures to ensure that basic definitions and methods of collection and analysis are consistent. Even the term 'population' requires qualification:

1. the **civilian** population of the UK, as might be guessed, excludes all armed forces; for some periods of this century, the merchant navy has also been excluded
2. the **total** population includes members of the UK armed forces whether at home or abroad
3. the **home** population is the *de facto* population, i.e. it includes armed forces of any nationality stationed in the UK and merchant seamen in home waters.

We have already mentioned the kind of discrepancies that can exist between Scotland and England; such differences are magnified when world-wide comparisons are to be made. For example, in Britain, a 'stillbirth' is defined to take place only after the 28th week of pregnancy; in the USA, the borderline is the 20th week. Efforts to harmonize this internationally have merely resulted in the definition of five stillbirth 'groups' which no one uses!

Likewise, changes in law may make it difficult to compare figures through time even within one country, e.g. introduction of the breathalyser in 1967 makes it very hard to relate road accident figures before and after this date. Tables published by government agencies usually carry footnotes indicating such changes.

6.4.1 Standard codes

One area where inconsistency can arise is in the coding of various attributes. Standard national and international codes exist for many of these but are subject to revision every few years. Some examples follow.

(a) Standard Industrial Classification

This codes industry of employment rather than occupation. Thus, a secretary working in a carpet factory (code 438) will be coded differently from the same grade of secretary working in a neighbouring household textiles factory (455). The code is an EEC activity classification and the version which came into use around 1980 has 10 divisions, 60 classes, 222 groups and 334 activity headings giving finer and finer detail,

e.g. 3 = manufacture of metal goods, engineering and vehicles
34 = manufacture of electrical and electronic equipment
344 = manufacture of telecommunications equipment
3443 = manufacture of radio and electronic capital goods

Changes of code are minimized, but new activities must be added to the list from time to time.

(b) International Standard Classification of Occupations

This refers to job rather than employing industry but is very long and complicated owing to world-wide coverage.

(c) Social class

This is an artificial five-category grouping based mainly on occupation of head of household. More than half of the gainfully employed population are in the very heterogeneous Class III but at least the coding is seldom changed. The social class concept has also been expanded into 17 **socioeconomic groups** with similar arbitrary divisions.

(d) International Statistical Classification of Diseases (usually abbreviated as ISCD or ICD)

This is a code for diseases and injuries, produced by the World Health Organization and revised approximately every ten years. It is basically a 3-digit code with an optional decimal figure for greater detail. Thus 427 is 'symptomatic heart disease' and 427.1 is 'left ventricular failure'. The codes in the range 800 to 999 relate to accident injuries, e.g. 845 is 'accident involving spacecraft'. Of course, such fine detail is seldom required and coarse groupings are usually used for summary, but even these are sometimes disturbed by coding changes. Psychiatric problems tend to be the most subject to such revision.

To reiterate, a comparison is invalid if we are not comparing like with like. Fallacious arguments in data interpretation are frequently the result of overlooking this. Beware!

6.5 OFFICIAL STATISTICS IN THE DEVELOPING COUNTRIES

Clearly, the problems of collecting and interpreting official figures for so-called Third World countries are enormous; reliability and consistency are particularly questionable. Since the Second World War, the United Nations Statistical Office has done much to stimulate a global view of economic and social problems particularly in relation to human fertility but the responsibility for gathering the required information is left with individual countries, the majority of which do not have resources of trained manpower to accomplish the task themselves.

The Statistical Office of the European Community is currently planning a 'twinning' scheme between certain African centres and selected European institutions in an attempt to remedy this lack. Skilled personnel are sometimes seconded from the West but there is a danger of imposing a Western framework that is totally unsuitable for societies where agriculture rather than industry is the mainstay of the community. For instance, in large areas of Africa and Asia, the concept of the nuclear family and therefore of the household is relatively meaningless. Likewise, 'employment' in the sense of earning money may have little relevance in a peasant society where people's needs are met by the labour of their own hands. Insistence on the use of such terms can introduce great problems of consistency when international comparisons are attempted. At the same time, international standards may demand tabular formats (for example, involving age groups) that are quite unsuitable for domestic planning.

In general, illiteracy and lack of motivation militate against reliability of figures from Third World countries. Unsatisfactory communications between capital city and enumeration districts may lead to misunderstanding, exacerbated by language and dialect difficulties as local enumerators proceed into the remoter areas. Even 'age' can cause problems in areas of the world where birthdays are not celebrated and, indeed, there is often a vague response to any question involving the passage of time. In some cultural environments, information is hard to obtain simply because bad luck is believed to follow upon its disclosure.

Even the simple head counting involved in a census can be difficult and subject to international inconsistencies. The distinction between *de jure* counting (practised in countries once in the French or Belgian sphere of influence) and *de facto* enumeration (the usual method in one-time British colonies) can lead to wildly differing estimates in areas with a high proportion of nomads.

Vital registration, even where it is theoretically compulsory, is also fraught with difficulties. For instance, there is good evidence that in Ghana in the 1960s only about 22% of births and 17% of deaths were being registered. This low coverage is, of course, largely due to the considerable distance of much of the population from the nearest registration office. A lengthy delay in the reporting of a birth increases the chance that the newborn child may die with no registration of either the birth or death; the national birth and death rates will obviously be underestimated in such circumstances. Naturally, the situation is constantly improving but unfortunately that also means that figures are not comparable over time.

Finally, it has been claimed (as much by statisticians from developing countries as by Western commentators) that official figures from the Third World are often considerably manipulated by the governments producing them. Sometimes this is done to present a biased picture of progress for domestic approval; sometimes the opposite is needed to support a request for foreign aid. Even within a country with ethnic and cultural divisions, false regional statistics may be presented to support claims for greater national representation, increased central funding, and so on. Interference with the true facts can be achieved by altering the usual format of tables, by selecting a favourable publication date and by adding explanatory notes that create a false impression. Political bias may also be introduced in, for example, the appointment of the head of the government statistical service or the allocation of research funding to favoured topics such as poverty. The cynical reader will suggest that these tactics are not peculiar to developing countries.

SUMMARY

Figures published by government agencies are freely available but the source, reliability and consistency of the data must be considered critically. Variations of definitions and standard coding schemes across national boundaries or through time are a particular problem. Quality of data in developing countries also presents special difficulties.

EXERCISES

1. Write an essay discussing the most recent census in your country. (For a 'developed' country, this will probably concentrate on the questions asked in the census form; for other parts of the world the method of conducting the census may be of more interest.)
2. Find out as much as you can about any one of the special UK sample surveys mentioned in Section 6.1.3; the government publications section of a moderate-sized library should provide most of what you need. Summarize the main features of the survey.
3. Try to track down more details of either the EEC Standard Industrial Classification or the International Statistical Classification of Diseases than are given in Section 6.4.1. (You may be lucky enough to have a sight of the complete listings or you may have to scan EEC or World Health Organization publications for information.)

7
Delving in tables

7.1 INTRODUCTION

Many readers of this book will have become involved with numbers in the belief that they are 'no good at English'. This is unfortunate since one of the most important facets of the use of data for scientific research, corporate management or political debate is the communication of the information contained therein to those who will say that they are 'no good at figures'! Even a table of results can be frightening, confusing or at least boring to many people unless a summary interpretation of the content is made available. Clearly a certain amount of skill is involved in the discussion of uncertainties and assumptions in detail sufficient to demonstrate objectivity and yet maintain a clear and convincing line of argument.

Because the presentation of results in lucid and effective language is so important, the qualifying examinations of many professional bodies explore candidates' aptitude for this and some employers (the Civil Service, for example) use similar tests in the screening of job applicants. The purpose of this chapter is to give some hints on how to comment on tables such as might be found in government summary publications. Masterly examples of this skill are to be found in the UK Government publication *Social Trends* and, though usually rather ambitious expositions, are certainly worthy of study and imitation.

In considering a government summary table, it must first be recognized that the raw data were probably collected with no specific use in mind and therefore, almost inevitably, too much detail is given or the format is unsuitable for the purpose in hand. We consider first how to reorganize a table to highlight the main features of interest. In particular, we go beyond the introduction to the tabular summary given in Chapter 2 to consider how the presentation of a table can be manipulated to reveal the information content to best effect.

7.2 TABULAR PRESENTATION

The first suggestion is to round the figures to two *effective* digits since psychologists have shown that the short-term memory need for scanning tables is limited to two digits. The mental subtractions and divisions necessary for examining the content of a table are thereby facilitated. This 'de-focussing' therefore helps to reveal any real patterns although, of course, it would have been unacceptable in the original table, the function of which was to provide an accurate record.

By **effective digits** we mean those that vary in the list of numbers under consideration. Thus, entries such as

870, 1076, 1521, 2774, 2163

having four effective digits would become

900, 1100, 1500, 2800, 2200

but

1638, 1797, 1025, 1672

having only three effective digits would be reduced to

1640, 1800, 1030, 1670

On the other hand, rounding must not become an inflexible rule, since numbers like

91, 107, 98, 103

would be considerably perturbed by such treatment. Naturally, after rounding, marginal or subtotals may be only approximately the sum of their components, but this can be explained in a footnote if necessary.

A phenomenon resulting only partly from use of the above technique is that figures can best be compared when arranged in columns since the eye can then relatively easily ignore unvarying digits and the brain can perform subtractions or divisions more readily. Long column headings must be abbreviated or made multi-line so that the eye travel for lateral comparisons is not excessive and to avoid the need to present a table side-on.

Another beneficial trick is to re-order the rows and columns of a table in descending order of marginal totals. This, again, aids the process of mental subtraction. On the other hand, if several similar tables are to be compared, the same format must naturally be used for all of them. Likewise, if past convention has made a particular layout familiar (such as time increasing downwards or from left to right), a change should be considered only if a clear advantage is to be gained. Whatever pattern is adopted, the rag-bag categories of 'miscellaneous', 'others' and 'not known' should appear as the bottom or right-most categories of rows and columns.

Where the row and column classifications provide a breakdown of a single population, expressing the body of the table as percentages of one or other marginal total may be helpful. But, if the information is derived from several populations (e.g. different years), the analagous technique involves setting a base year figure to 100 and scaling the figures for succeeding years relative to this. (The use of indices is discussed more fully in the next chapter.) Since we are generally concerned to detect deviations from average, it can be useful to show the row and column averages where these make sense. This gives a starting point from which to look for patterns of discrepancy in the subcategories.

Lastly, careful use of space in the table layout may go a long way towards helpful guiding of the eye. As an alternative to ruled horizontal lines, any subtotals in columns can be offset to the right a little to assist the reader to focus on them or discard them as required. In printed tables, even the comma used to subdivide large numbers into groups of three digits is often replaced with advantage by a small gap.

7.3 WRITING ABOUT TABLES

There can be no hard-and-fast rules about how to elicit from a table or group of tables the main points of interest; it is possible to offer only some general guidelines. As a starting point, the methods of tabular reformulation discussed in the last section may form part of an initial familiarization phase. Notes can be made at this stage and possibly some rough diagrams drawn which may or may not be incorporated into the final document. Once a reasonably complete distillation has been achieved, writing can begin.

The first consideration must be the nature of the intended audience since this will determine the level of technical jargon (statistical or otherwise) that may be permitted, the format and length of the exposition and the style of writing to be adopted. For example, whether or not subheadings (possibly numbered) should be used will be partly related to length; they will certainly be of benefit in any text longer than 500 words. Similarly, the technical expert will be content to read a much more dry but precise report than the apathetic 'man in the street'.

Occasionally, it can be worth while to vary the style. In this book, for example, there are passages (e.g. Section 5.5) where a somewhat pompous style is used to present one side of a possibly debatable viewpoint. By way of contrast, a more informal approach (e.g. Section 1.3) is used with the aim of encouraging readers through a potentially offputting topic. Sometimes an intentionally condensed presentation (e.g. Section 17.5) may 'make the reader think', but such a tactic can be dangerously optimistic. (You now know, dear reader, how to view the obscure and cryptic sections of this text!) In any event, do not descend to the clichés of politician and journalist for whom all changes are 'dramatic', all majorities are 'vast' and all proportions are 'mammoth'.

Since there is no agency to arbitrate on what is correct and what is not in the use of English, the language is free to mutate quite rapidly. Nevertheless, current grammatical freedom is unpalatable to many educated in a more rigid era. It will generally be politic to avoid the antipathy of these readers by eschewing forms such as 'different to', 'less people' and 'the most numerous sex'. However, publisher's spelling conventions (e.g. -ize) may have to be accepted.

It will almost certainly be useful to begin your report with some comments on the background to the tabulated information. For example, the three aspects of source, reliability and consistency of data discussed in Chapter 6 may provide a useful framework. Footnotes to a table may indicate inconsistencies (over years, for example) due to changes of definition. Whatever the overall reliability of the figures, some categories may be less reliable than others due to particular circumstances or simply because the relevant subsamples are very small. A comment on the source might extend to whether or not the data are routinely collected and, if not, what was the motivation behind the particular study.

In presenting the main commentary, remember that readers will either have the original table before them or a number of tabular extracts presented by you to clarify points of interest. Thus it will generally not be useful to restate raw figures in the text particularly since comparisons are a vital component of the exercise and these are best expressed as percentages or appropriate rates. Do not be afraid to aggregate categories that are not (for your purposes) usefully differentiated but, on the other hand, mention parallel or inverse trends in subclasses if you feel these represent more than sampling variability.

While your commentary should naturally be unbiased, it is reasonable that it should include possible reasons for any patterns revealed. Such explanations could be reasonably well founded or might be hypothetical, in which case this should be made clear to the reader. It is inevitable that many social and medical phenomena relate to sex, age, social class and race (i.e. genetic or cultural) differences. The independence of events such as deaths from plane crashes is also important since this may relate to the appropriateness of any rate quoted. In addition, 'when?' and 'where?' are often important attributes of events as are fluctuating patterns throughout the day, week or year (e.g. road traffic accidents) and discontinuities across regional and national boundaries with varying degrees of legal or cultural similarity (e.g. divorce rates in England and Scotland).

Finally, it is quite in order to conclude with a brief mention of any potentially useful information that is lacking either because it is not available to you or because it has never been collected (but could be). Such auxiliary data might, for instance, confirm a suggested explanation or help to formulate a more appropriate rate for some event.

SUMMARY

The extraction and re-presentation of information from published tables is an important statistical activity. Practice develops the skill but a few guidelines provide possible approaches to the task.

EXERCISES

1. Choose a moderately extensive table from a government publication and summarize its main features in, say, 700–800 words. Include tabular restatement and/or graphical presentation if these are helpful.

2. Tables E7.1 and E7.2, reproduced from the *Digest of Welsh Statistics* (1986), gives

information on legal abortions. Write a report on these data using graphical representation where this is helpful and indicating any other figures that could throw further light on the situation.

Table E7.1 Legally induced abortions reported by hospitals in Wales Number

	1971	1976	1980	1981	1982	1983	1984	1985(a)
By marital status:								
Single	1147	1048	1372	1550	1593	1721	1800	2048
Married	2057	1570	1632	1666	1486	1449	1391	1394
Others (b)	306	320	483	461	508	439	484	534
Not stated	15	—	3	7	—	—	—	1
By age:								
Under 16	91	110	112	115	136	138	133	150
16–19	627	614	793	878	838	891	930	1016
20–34	1996	1589	1850	1950	1916	1880	1970	2149
35–44	780	595	700	714	669	685	616	648
45 and over	24	30	34	27	28	15	26	14
Not stated	7	—	1	—	—	—	—	—
By parity: (c)								
0	1117	1059	1369	1527	1542	1641	1707	1905
1	362	340	515	463	502	516	533	617
2	639	746	846	906	883	827	830	865
3	601	488	508	492	431	401	388	410
4	371	183	173	196	161	136	160	134
5 +	435	122	79	97	68	88	57	46
Not stated	—	—	—	3	—	—	—	—
By method:								
Vacuum aspiration	1964	2295	2591	3029	2917	3000	3070	3330
Dilation and curettage	114	54	55	45	16	65	71	26
Hysterotomy	978	101	38	25	19	21	12	13
Hysterectomy	65	23	20	8	7	1	5	4
Other surgical	..	..	..	3	7	9	8	4
Prostoglandins only	..	..	..	414	372	341	338	336
Prostoglandins and other	..	..	..	156	248	170	171	264
Other medical	..	..	..	4	1	2	—	—
Not stated	404	465	786	—	—	—	—	—
By sterilization:								
Sterilized	994	395	446	472	400	328	302	298
Not sterilized	2531	2543	3044	3212	3187	3281	3373	3679
By gestation time:								
Under 9 weeks	385	290	577	656	673	684	688	611
9 weeks and under 13 weeks	1917	1834	2008	2294	2245	2329	2374	2669
13 weeks and over	1176	812	872	725	669	596	613	697
Not stated	47	2	33	9	—	—	—	—
Total abortions	3525	2938	3490	3684	3587	3609	3675	3977

(a) Provisional figures. Notifications received up to 14 July 1986. Source: Welsh Office
(b) Widowed, divorced or separated.
(c) Number of previously born children.

Table E7.2 Legally induced abortions to Welsh residents, by place of operation

Number of operations	1971	1976	1981	1982	1983	1984	1985(a)
Wales	3674	2932	3662	3621	3477	3623	3829
England	1124	1933	2679	2481	2387	2614	2768
Total	4789	4865	6341	6102	5864	6237	6597
Rates (b)	9.4	9.2	11.2	10.6	10.1	10.7	11.2

(a) Provisional
(b) Per 1000 women aged 15–44

Source: Office of Population Censuses and Surveys

3. Table E7.3 comes from the September 1987 *Monthly Digest of Statistics*. (The June Census is a compulsory return from all farmers supplying information on crops grown, etc.; φ means figures provisional.) Describe the main features of the table. Find information on European Community land use and crop areas and make comparisons.

Table E7.3 Land use and crop areas[1]

	Area at the June Census (Thousand hectares)					
	1982	1983	1984	1985	1986	1987$^{\varphi}$
Total crops	5 072	5 027	5 154	5 224	5 239	5 268
Bare fallow	55	97	42	41	48	40
All grasses under five years old	1 859	1 846	1 794	1 796	1 723	1 699
All grasses five years old and over	5 097	5 107	5 105	5 019	5 077	5 108
Sole right rough grazing	4 984	4 927	4 895	4 872	4 829	4 773
All other land on agricultural holdings, including woodland	502	519	517	535	543	547
Total area on agricultural holdings	17 569	17 523	17 508	17 487	17 459	17 436
Common rough grazing (estimated)	1 214	1 212	1 212	1 216	1 216	1 216
Total agricultural area	18 783	18 735	18 720	18 703	18 676	18 652
Total area of the United Kingdom	24 088	24 088	24 088	24 085	24 086	. .
Crops						
Cereals (excluding maize) Total	4 030	3 961	4 036	4 015	4 024	3 940
Wheat	1 663	1 695	1 939	1 902	1 997	1922
Barley	2 222	2 143	1 978	1 965	1 916	1 836
Oats	129	108	106	133	97	100
Mixed corn	10	8	8	7	7	6
Rye	6	7	6	8	7	7
Rape grown for oilseed	174	222	269	296	299	391
Sugar beet, not for stockfeeding	204	199	199	205	205	203
Potatoes: Early crop	25	24	198}	191	178	178
Main crop	167	171				
Mainly fodder crops:						
Maize for threshing or stockfeeding	16	15	16	20	23	24
Field beans[2]	40	34	32	45	60	91
Turnips and swedes[3]	71	66	64	62	59	56
Fodder beet and mangolds[3]	5	5	8	12	13	13
Kale, cabbage, savoy, kohl rabi and rape	43	40	39	37	33	29
Peas harvested dry[4]	. .	29	56	92	91	119
Other crops for stockfeeding	31	25	22	20	20	12
Hops	6	6	5	5	4	4
Other crops, not for stockfeeding	6	7	6	11	18	11

(*Table contd overleaf*)

Table E7.3 (*Contd.*)

	Area at the June Census (Thousand hectares)					
	1982	1983	1984	1985	1986	1987$^{\varphi}$
Horticultural crops (excluding mushrooms) Total[5,6]	254	224	204	212	213	197
Brussels sprouts	13	11	11	11	11	..
Cabbage (all kinds)	13	12	13	12	13	..
Cauliflower and calabrese	14	13	12	12	13	..
Carrots	14	13	14	14	14	..
Parsnips	3	2	3	3	3	..
Turnips and swedes[7]	3	..	..	..	..	..
Beetroot	2	2	2	2	2	..
Onions salad } Onions dry bulb }	9	8	8	9	9	..
Beans broad } Beans runner and French }	12	11	10	11	10	..
Peas for harvesting dry[6]	27	18	..	..	..	..
Peas green for market } Peas green for processing }	56	47	46	53	52	..
Celery (field grown)	1	1	1	1	1	..
Lettuce (not under glass)	4	4	4	4	4	..
Other vegetables and mixed areas	9	10	10	12	12	..
Orchards: commercial	40	38	37	36	36	..
non-commercial	3	2	2	2	2	..
Small fruit[8]	18	17	16	16	15	..
Hardy nursery stock	7	7	7	7	7	..
Bulbs grown in the open	5	4	4	4	4	..
Other flowers grown in the open	1	1	1	1	1	..
Area under glass or plastic covered structures	2	2	2	2	2	..

Sources: Agricultural Departments

1. Figures include estimates for minor holdings in England and Wales but not for Scotland and Northern Ireland. For further details refer to the *Supplement of Definitions*.
2. Prior to 1986 collected as 'Beans for stock feeding' in England and Wales.
3. Fodder beet in Northern Ireland is included with turnips and swedes. In Scotland fodder beet was collected separately for the first time in 1986. It was previously included with turnips and swedes.
4. Includes 'Peas for harvesting dry for both human consumption and stockfeeding' from 1984 onwards.
5. Figures relate to land usage at 1 June and are not necessarily good indicators or production as for some crops more than one crop may be obtained in each season or a crop may overlap two seasons.
6. Following a change of definition in 1986 'Horticultural crops' now excludes 'Peas for harvesting dry for human consumption' as well as 'Mushrooms'. The data from 1984 reflect this change.
7. Included with other vegetables from 1983 onwards.
8. Excludes area of small fruit grown under orchard trees in England and Wales which is returned as orchards.

8
Changes with time – index numbers

8.1 INTRODUCTION

The use of indicators of current economic performance relative to past performance has become more and more widespread during the last 50 years. Government decisions, national and international, negotiations with producers over prices and with unions over wages, etc., depend increasingly on such measures. Typically an **index number** indicates the **current** position of prices (e.g. UK index of retail prices) or quantities (e.g. UK index of industrial production) of a list of commodities relative to a specified **base** point (though the principles can apply in other situations). (Strictly speaking, 'current' is a jargon term that need not indicate present time, as the example in the next section shows.)

Most forms of index number are essentially weighted means such as were described in Section 3.1.5. Let us see how this comes about.

8.2 SIMPLE INDICES

Suppose we wish to compare someone's purchases of first- and second-class postage stamps in 1981 (current) with the corresponding figures for 1976 (base). We may be interested in the price comparison or in the quantity of stamps bought. The relevant figures might be as shown in Table 8.1.

We could compare the changes for each postal rate using the columns headed 'price relative' and 'quantity relative'. Thus the price relative for first-class stamps is $100 \times 14/8.5 = 164.7$, indicating an increase in price of 64.7%. The quantity relative for second-class stamps is 140.0, indicating an increase in numbers bought of 40%.

To compare the costs over both postal rates, there would appear to be two possibilities:

1. We could average the relatives. This would give us overall price and quantity relatives of $341.6/2 = 170.8$ and $260/2 = 130.0$ respectively;
2. We could calculate the relative of the averages. The average price moved from $15.0/2 = 7.5$ to $25.5/2 = 12.75$, giving a relative of 170.0; the corresponding quantity relative is 130.8.

Table 8.1 Prices and quantities of first- and second-class stamps in 1976 and 1981

	Base year (B = 1976)			Current year (C = 1981)		
	Price p_B	Quantity q_B	Value $= p_B q_B$	Price p_C	Quantity q_C	Value $= p_C q_C$
First class	8.5	60	510	14.0	72	1008
Second class	6.5	70	455	11.5	98	1127
Totals	15.0	130	965	25.5	170	2135

	Price relative $= 100 \times p_C/p_B$	Quantity relative $= 100 \times q_C/q_B$	Values of 1981 quantities at 1976 prices $= p_B q_C$	Values of 1976 quantities at 1981 prices $= p_C q_B$
First class	164.7	120	612	840
Second class	176.9	140	637	805
Totals	341.6	260	1249	1645

It may be thought disturbing that the two methods give different answers; however, the real argument against *both* is clearer when they are applied in a situation where the purchases are of different commodities, e.g. coal (in tons), electricity (in kilowatt hours), matches (in boxes of 50). In such a situation, average price depends on units of measurement and average quantity is meaningless; hence method 2 above is useless. At least method 1 is independent of the units of measurement (providing they are the same in both years) but the equal weight that attaches to all commodities can give a misleading figure – if the price of matches falls by 50% and the price of coal rises by 50% the combined index indicates no change. This seems unreasonable in a household budget.

8.3 WEIGHTED INDICES

8.3.1 The Laspeyres index

We must therefore adapt method 1 by **weighting** the commodities, i.e. taking a weighted mean of the price or quantity relatives (see Section 3.1.5). What weights should we use? Clearly, our view of the importance of the 64.7% increase in the price of first-class stamps or the 40% increase in the purchase of second class stamps depends on the *value* involved. Thus it could be appropriate to calculate a weighted mean of the relatives using the base values, $p_B q_B$, as weights; i.e. for the price index,

$$100 \times \frac{(14/8.5) \times 510 + (11.5/6.5) \times 455}{510 + 455} = 100 \times \frac{1645}{965} = 170.5$$

Notice that the formula we have used is

$$100 \times \frac{\sum(p_C/p_B) \times (p_B q_B)}{\sum(p_B q_B)}$$

which reduces to

$$100 \times \frac{\sum(p_C q_B)}{\sum(p_B q_B)}$$

and this is just the relative of the averages of prices weighted by base quantities, i.e. equivalent to a weighted version of method 2. The result can be simply expressed as

$$100 \times \frac{\text{total value of base quantities at current prices}}{\text{total value of base quantities at base prices}}$$

Thus, prices are compared as a ratio of values of base quantities.

Similarly, quantities are compared as a ratio of values at base prices, i.e.

$$100 \times \frac{\sum(q_C/q_B) \times (p_B q_B)}{\sum(p_B q_B)} = 100 \times \frac{\sum(p_B q_C)}{\sum(p_B q_B)}$$

$$= 100 \times \frac{\text{total value of current quantities at base prices}}{\text{total value of base quantities at base prices}}$$

$$= 100 \times 1249/965 = 129.4$$

These index numbers, using base period value weights, were invented by the German economist Laspeyres.

8.3.2 The Paasche index

We could equally well construct an index in a way suggested by Paasche. The value weights used are hypothetical and, for price comparisons, represent the value if current quantities had applied in the base year. For quantity comparisons, the hypothetical value is that which would have held if current prices had existed in the base year. The calculations are as follows:

$$\text{Paasche price index} = 100 \times \frac{\sum(p_C/p_B) \times (p_B q_C)}{\sum(p_B q_C)}$$

$$= 100 \times \frac{\sum(p_C q_C)}{\sum(p_B q_C)}$$

$$= 100 \times \frac{\text{total value of current quantities at current prices}}{\text{total value of current quantities at base prices}}$$

$$= 100 \times 2135/1249 = 170.9$$

$$\text{Paasche quantity index} = 100 \times \frac{\sum(q_C/q_B) \times (p_C q_B)}{\sum(p_C q_B)}$$

$$= 100 \times \frac{\sum(p_C q_C)}{\sum(p_C q_B)}$$

$$= 100 \times \frac{\text{total value of current quantities at current prices}}{\text{total value of base quantities at current prices}}$$

$$= 100 \times 2135/1645 = 129.8$$

8.4 COMMENTS ON THE LASPEYRES (L) AND PAASCHE (P) INDICES

The two indices appear to have equal validity in measuring the relative change in price or quantity levels between two periods. In general L and P are not equal and it is therefore reasonable to ask if there are theoretical reasons for favouring one over the other.

1. Neither index takes account of the fact that, in a free market, people adjust their buying habits to get greatest value, i.e. if prices change, quantities bought may also change. Thus, by insisting on a constant quantity pattern, the value of base quantities at current prices overestimates the real current value and hence L overestimates the real effect of price changes. By a similar argument, P underestimates the effect. Nevertheless it is still possible to find L less than P, as in our example.
2. For L, the weights have to be established only once. This speeds the production of the index number where complex commodity lists are involved so that there is less delay in publication. On the other hand, the base weights can quickly become irrelevant in a dynamic economy (see 1 above). For P, the weights have to be recalculated for each current year and, again, current patterns might have been unreasonable in the base year.

8.5 CHANGING THE BASE

Clearly the base point must be one at which reasonable market stability existed. For example, we should not use a month in which there was a generous round of price increases since that will give an impression of little change in later months. Such stability may be difficult to achieve for all commodities in a complex list.

Sooner or later the base must be changed and the following example demonstrates the difficulty of relating the new index to the old base for long-term comparisons.

Now, the first- and second-class stamp prices in 1986 were 17p and 12p respectively, so we have:

	p_{86}	$p_{86}q_{81}$	$p_{86}q_{76}$
First class	17	1224	1020
Second class	12	1176	840
	Totals	2400	1860

Then L for price in 1986 relative to 1981 is

$$100 \times 2400/2135 = 112.4$$

and we have already calculated L for 1981 relative to 1976 to be 170.5. Hence we are suggesting that L for 1986 relative to 1976 might equal $100 \times 1.124 \times 1.705 = 191.6$

Unfortunately, L for 1986 relative to 1976 equals

$$100 \times 1860/965 = 192.7$$

so that the conversion is incorrect for L (and likewise for P). It would be valid only if

1. all quantities were to alter by the same factor from 1976 to 1981, or
2. we use fixed 'standard' quantities that belong neither to the base year nor to the current year but are decided on some other basis.

Nevertheless, such a link is often made even when this is not the case. It is of particular relevance in the construction of **chain-linked index numbers** where the comparison is made as a sequence of year-to-year (or month-to-month, etc.) indices using either of the index numbers described. This facilitates frequent changes in the commodity list and, as in the case above, the error is not usually very great.

8.6 UK INDEX OF RETAIL PRICES

This is designed to show month-to-month changes in average prices of goods and services purchased by households. It is not really a cost of living index since many goods of a luxury nature are included and several items (e.g. income tax, insurance premiums, etc.) for which it is hard to identify 'units' of sale are excluded.

Until 1987, the index used 11 major groups: food, alcoholic drink, tobacco, housing, fuel and light, durable household goods, clothing and footwear, transport and vehicles, miscellaneous goods, services, and meals consumed outside the home.

At the start of 1987, a new 14-group system was introduced. Alcoholic drink, tobacco, housing, fuel and light, and clothing and footwear are as before, but the remaining groups have been restructured as food, catering, household goods,

Table 8.2 Weights for index of retail prices

	1956–62	1963	1968	1972	1976	1980	1982	1984	1986
Food	350	319	263	251	228	214	206	201	185
Alcohol	71	63	63	66	81	82	77	75	82
Tobacco	80	77	66	53	46	40	41	36	40
Housing	87	104	121	121	112	124	144	149	153
Fuel & light	55	63	62	60	56	59	62	65	62
Durables	66	64	59	58	75	69	64	69	63
Clothing etc.	106	98	89	89	84	84	77	70	75
Transport	68	93	120	139	140	151	154	158	157
Misc.	59	63	60	65	74	74	72	76	81
Services	58	56	56	52	57	62	65	65	58
Meals out	—	—	41	46	47	41	38	36	44

Source: Table 18.6 of *Annual Abstract of Statistics*, 1987, (Department of Employment) and Table 18.1 of *Monthly Digest of Statistics*, January, 1987.

household services, personal items, motoring expenditure, fares etc., leisure goods, and leisure services. Naturally, these inconsistencies do not affect the comparability of the 'all items' index over time and, in part, the changes were introduced to conform more closely with international classifications. Each group is divided into a number of sections and, for each section, representative goods and services have prices recorded. These are transaction prices, ignoring hire purchase charges, 'co-op' dividends and discounts not given to all customers. The prices are collected by postal contact with manufacturers or visits to retailers in a representative list of towns.

The weights used were originally calculated from an expenditure survey of 11 638 households in 1953–54. From 1962 to 1974 the weights were recalculated every February from the results of the latest three Family Expenditure Surveys. Since 1975, the index has been current weighted by expenditure in the latest available year. Table 8.2 shows how the weights have altered over the years. (Notice that the weights in each column sum to 1000.) The figures reveal fairly consistent reductions in the weights for food and tobacco, and an increase (levelling off) for transport and an unsteady (!) fluctuation for alcohol.

The prices are usually collected on the Tuesday closest to the 15th of the month and the index is published in the third week of the following month in various economic journals.

8.7 UK INDEX OF INDUSTRIAL PRODUCTION

This is a quantity index designed to measure monthly changes in the volume of UK industrial production. The definition of 'industrial' is a little arbitrary – mining, quarrying, gas, electricity and water are included, but service industries, agriculture and fishing are not. The output is recorded from both private- and public-sector firms whether the goods are for domestic consumption, export or the armed forces.

The index incorporates 880 items showing the difference between output of products and input of materials, etc., integrated by the Laspeyres base-weighted method. Where values are supplied by a firm, these have to be adjusted to the price levels for the base period, otherwise the index would reflect value rather than quantity changes. Where production extends over a long period as in construction industries and shipbuilding, measurement of work in progress has to be achieved by other means.

To take account of different lengths of calendar months, a 'standard month' is used, but, even so, public holidays, particularly at Christmas, can distort the index. The main problem for this index is the considerable delay before publication. It can take as much as six months for all required figures to be available and it is usual to issue a provisional index six to seven weeks after the end of the month in question and revise this later.

SUMMARY

Indices are available that can be used to compare prices (or, less often, quantities) at two time points. Different commodities are combined as weighted means of the relative change ratios.

EXERCISES

1. The table below gives the values (in £million) of gross domestic fixed capital formation in three sectors of UK industry at both current and 1980 prices for three separate years.

	At prices current in			At 1980 prices		
	1976 (a)	1980 (b)	1986 (c)	1976 (d)	1980 (e)	1986 (f)
Private sector	14 389	29 409	51 376	24 965	29 409	36 159
General government	5 422	5 499	7 296	9 117	5 499	6 112
Public corporations	4 693	6 653	5 555	8 006	6 653	4 275

Source: Tables 12.1 and 12.2 of UK National Accounts, HMSO (1987)

(a) Explain why the entries in columns (b) and (e) are identical and why the entries in columns (a) and (d) differ.
(b) Explain how to calculate from these figures base-weighted (Laspeyres) price index numbers with 1980 as base year.
(c) For the three sectors combined and with 1980 as base year, calculate current-weighted (Paasche) price index numbers for 1976 and for 1986.

2. Table E8.1 comes from the 1987 *Digest of UK Energy Statistics* (Table 63). Discuss the patterns revealed by the data, making use of the techniques discussed in this chapter. (If you have access to the original publication, you might find the explanatory notes on p. 90 of interest.)

Table E8.1 Consumers' expenditure on energy (1)
United Kingdom

Prices and values
£ Million

	1982	1983	1984	1985	1986
At current market prices					
Coal (2)	728	738	659	908	786
Coke	101	102	106	125	112
Gas	3 063	3 530	3 671	4 046	4 372
Electricity	4 264	4 450	4 551	4 860	5 196
Petroleum products (3)	528	566	575	704	667
All fuel and power (4)	8 696	9 399	9 575	10 657	11 148
Motor spirit and lubricating oil	6 331	6 913	7 450	7 972	7 326
Total energy products (5)	15 027	16 312	17 025	18 629	18 474
Total consumers' expenditure	167 382	182 600	195 341	213 235	231 632
Revalued at 1980 prices					
Coal (2)	574	550	441	578	487
Coke	80	77	74	81	71
Gas	1 945	1 985	2 000	2 121	2 241
Electricity	3 237	3 244	3 278	3 393	3 547
Petroleum products (3)	363	351	346	382	374
All fuel and power (4)	6 210	6 218	6 149	6 565	6 730
Motor spirit and lubricating oil	4 887	4 985	5 180	5 275	5 587
Total energy products (5)	11 097	11 203	11 329	11 840	12 317
Total consumers' expenditure	138 285	143 610	146 667	151 961	159 165

(1) These figures are based on National Income and Expenditure in the Fourth Quarter of 1986 – an article in the April 1987 issue of *Economic Trends* published by the Central Statistical Office. The figures exclude business expenditure.
(2) Including some manufactured fuel.
(3) Excluding motor spirit and lubricating oil.
(4) Including an estimate for wood.
(5) Quarterly data on energy expenditure is shown in the Central Statistical Office's *Monthly Digest of Statistics*.

9
Demography – introduction

9.1 THE IMPORTANCE OF DEMOGRAPHIC STUDIES

Demography is concerned with description of the size, composition and structure of populations (usually human) and *how these change with time*. Thus, we are interested in patterns of births, deaths, marriages and migration (event counts) and the distribution of attributes such as age, sex, social class, family size, etc. (head counts). The data for such studies come from censuses, vital registration, etc.

This kind of description will be of interest to, for example, geographers (spatial distribution of populations), health service and other planners (future provision of maternity, geriatric, etc., services), sociologists (what decides family size?), economists (changes in per capita income), actuaries (pension arrangements), and many others.

One main problem is that of the delayed effects of population fluctuation. A 'baby boom' now could provide a large number of teachers in 20 years' time; but if in 15 years' time, the birth rate were to be low, there would be insufficient children for these teachers to teach (unless pupil/teacher ratios were to be changed).

Future predictions are not all equally dependable. For instance, on the basis of last year's births, we can make a fairly accurate assessment of the number of first-year secondary school places required in 11 or 12 years' time, because this depends mainly on fairly stable child mortality patterns. But we cannot estimate so reliably how many of these will seek university places in 17 or 18 years' time since this will be affected so much by changeable social and economic factors.

Demographic techniques are also of some use in ecological surveillance where there is interest in the breeding stability of a colony of birds or animals or where concern exists over culling programmes or the depletion of fish stocks. Indeed, some of the methods are of greater validity in these circumstances due to the relative brevity of animal lifetimes and freedom from the rapid temporal changes that human society brings upon itself.

Likewise, many of the methods are useful in manpower studies where the population under investigation is a large workforce, recruitment and leaving (for whatever reason) are the analogues of birth and death respectively, and we are interested in the stability of the age/sex structure or length of service patterns and the prediction of natural wastage.

9.2 DYNAMICS OF POPULATION CHANGE

The year-to-year change in total number in a population depends on two pairs of components:

1. natural increase or decrease, i.e. the difference between births (B) and deaths (D)
2. net migration, i.e. the difference between immigration (I) and emigration (E); this can be negative during periods of restricted immigration.

Thus, the population in year t, P_t, is related to that in the previous year by

$$P_t = P_{t-1} + (B - D) + (I - E)$$

Ignoring these components for the present, we can simply consider a population growing at say, 0.5% per annum (1 in 200 – roughly the UK rate). Then

$$\begin{aligned} P_{1986} &= P_{1985} \times 1.005 \\ &= P_{1984} \times (1.005)^2 \\ &\text{and so on} \end{aligned}$$

so that, in general,

$$P_t = P_0 \times (1.005)^t$$

where P_0 was the population size in the starting year. (This is just the formula for capital growth when compound interest is applied.)

How long will it take for the population to double, i.e. when do we have to feed twice as many people? We require to find t such that $P_t = 2P_0$ and the answer turns out to be about 140 years. We have to make the very unreasonable assumption that the growth rate stays constant during all those 140 years. In some countries, the present growth rate is about 2.5% and the time taken for such populations to double (under the same assumption) is therefore only 28 years!

We shall return to population projection later.

9.3 DEMOGRAPHIC TRANSITION

Before proceeding further, let us consider briefly one aspect of the way in which human populations have been observed to grow.

Little accurate information is available on population sizes in the middle ages but such evidence as we do have for the UK suggests that there was a six- or sevenfold increase between AD 1100 and 1780, with alternating lengthy periods of rapid and slow growth. These fluctuations were the direct results of famines, plagues, climatic variations and, to a lesser extent, international and civil strife. From about 1780 onwards, the population grew so steadily and so rapidly that by 1950 the number of inhabitants had again increased by seven times. With few differences, this pattern was repeated throughout Europe. Such a relatively abrupt change in the rate of increase of population size has become known as **demographic transition.**

This phenomenon was accompanied by a general movement from rural areas to towns and a gradual decline in the death rate, followed by a fall in the birth rate. The lag in birth rate decrease behind death rate decrease is as characteristic of developing countries today. It takes some years for the populace to realize that couples need not conceive so many children merely to ensure the survival of two or three. It is largely this lag that initiates demographic transition but other factors undoubtedly complicate the picture.

For instance, there may be economic and social influences. The Industrial Revolution in Europe brought about an environment with a considerable element of child labour, making large families advantageous. At about the same time, in England and Wales, poor relief was, for the first time, made dependent on family size (partly to alleviate the effects of the Napoleonic Wars). Other changes meant that agricultural workers became less dependent on their employers for lodgings and could therefore marry younger.

Coupled with these economic and social changes, there was a vast improvement in the health of the community. This was in part because of better diet (due to new crops introduced from abroad and more reliable agricultural methods). But, above all, better water supplies and the construction of sewers meant improved public hygiene. At a personal level, the availability of cheap soap and the gradual change from wool to more easily washed cotton clothes did much to make people cleaner.

However, some of the health improvement came about through natural reduction in disease. For example, bubonic plague had largely disappeared either because rats developed an immunity to the bacillus or due to a change in the flea transmitters. Likewise, immunity to typhoid and smallpox seems to have increased spontaneously in the human population. Vaccination, of course, finally eliminated smallpox and there appeared other 'drugs' such as mercury to treat syphilis and quinine to prevent malaria (though greater drainage of bogs was as much responsible for control of the latter). Increase in the use of isolation hospitals and much improved midwifery techniques did much to lower infant and maternal mortality and this is obviously a key aspect of demographic transition.

The relative importance of these factors in different countries (and even in the UK) remains the subject of much debate among economic and social historians. That the phenomenon of demographic transition exists is unquestionable though, nowadays, the timescale of change is so condensed that the metamorphosis of a developing country may appear superficially different from that of eighteenth-century Europe.

9.4 AGE–SEX PROFILES

A pictorial representation of the age–sex structure of a population at a particular time is provided by an **age–sex profile** or **population pyramid.** Figures 9.1 and 9.2

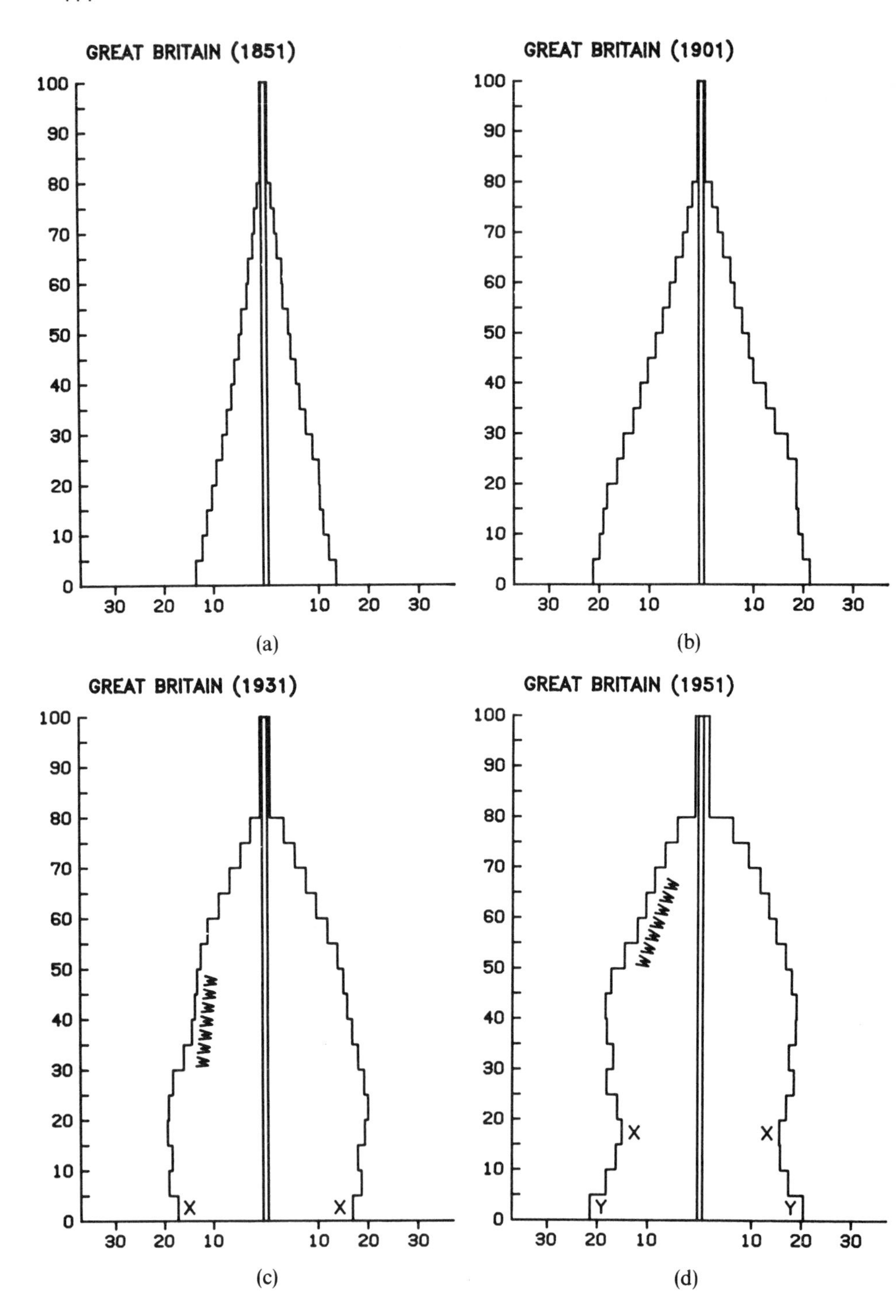
GREAT BRITAIN (1851)
GREAT BRITAIN (1901)
GREAT BRITAIN (1931)
GREAT BRITAIN (1951)
100
90
80
70
60
50
40
30
20
10
0
30 20 10 10 20 30
WWWWWW
WWWWWWW
X
Y
(a)
(b)
(c)
(d)

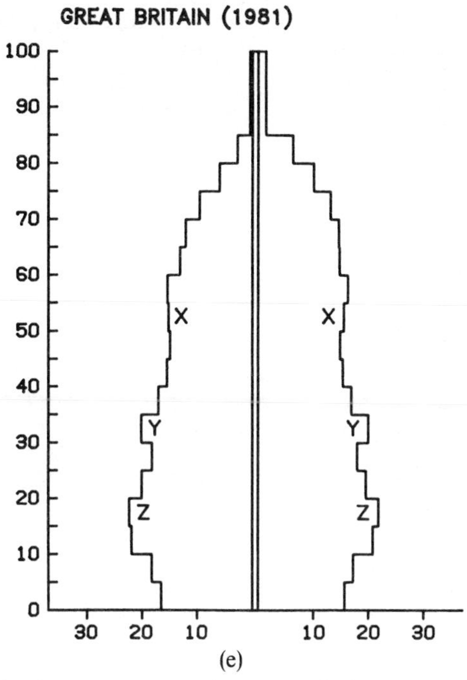

(e)

Fig. 9.1 Age–sex profiles for Great Britain over a period of 130 years. Horizontal axes show absolute numbers (thousands) in population; vertical axes show age in years. (a) Total population = 19 million. (b) Total population = 37 million. (c) Total population = 45 million. (d) Total population = 49 million. (e) Total population = 54 million. See text for further explanation.

give some examples. In drawing such profiles, the male part is conventionally drawn on the left and an arbitrary decision must be made on where to terminate the open-ended top age class (say 100 years). Five-year age groups are usual but, of course, other groupings could be used.

The diagram can portray actual *numbers* in the age–sex classes so that differences in total population size are apparent. Figure 9.1 shows the pattern of demographic transition for Great Britain in 'snapshots' from 1851 until the present day. Notice that the pyramid shape is typical only during a period when medical care is deficient; birth and death rates are then high with probability of death being relatively constant at all ages. As medical care improves, a much more 'beehive' shaped profile becomes the standard pattern with death being fairly unlikely before the age of about 60, at which point the remaining scourges of cancer and heart disease maintain the pyramid form for the older age groups. By 1901, the population had grown considerably but increased use of birth control was narrowing the lower part of the pyramid. Notice in the 1931 profile the effect of the First World War (WWWWW) on the male population and the drop in births (XX) due to the start of the 1925–35 depression. The remaining diagrams of

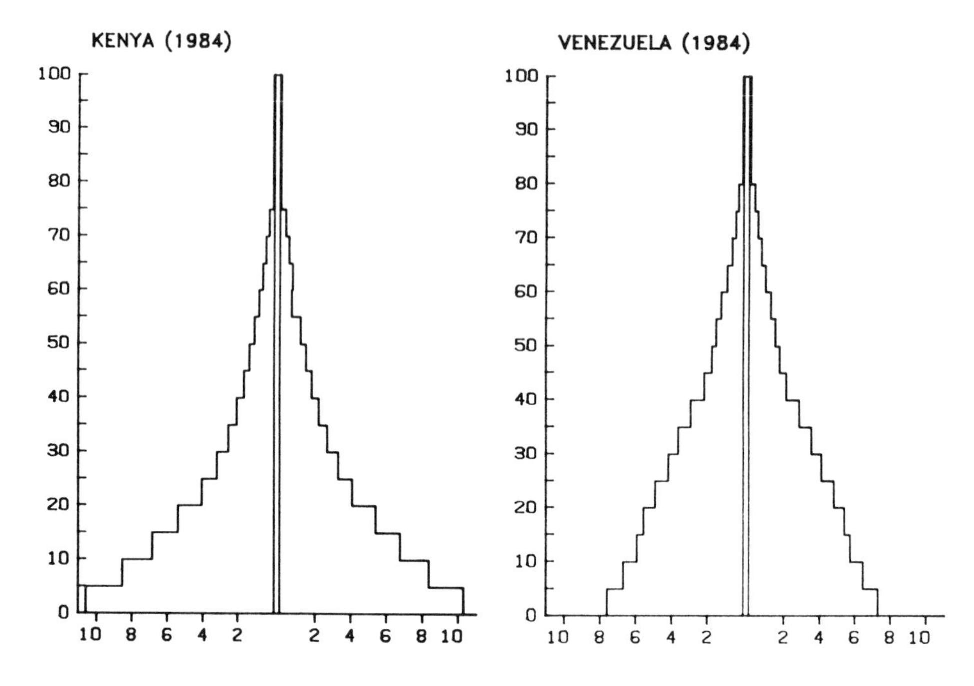
KENYA (1984)
VENEZUELA (1984)
100
90
80
70
60
50
40
30
20
10
0
10 8 6 4 2 2 4 6 8 10

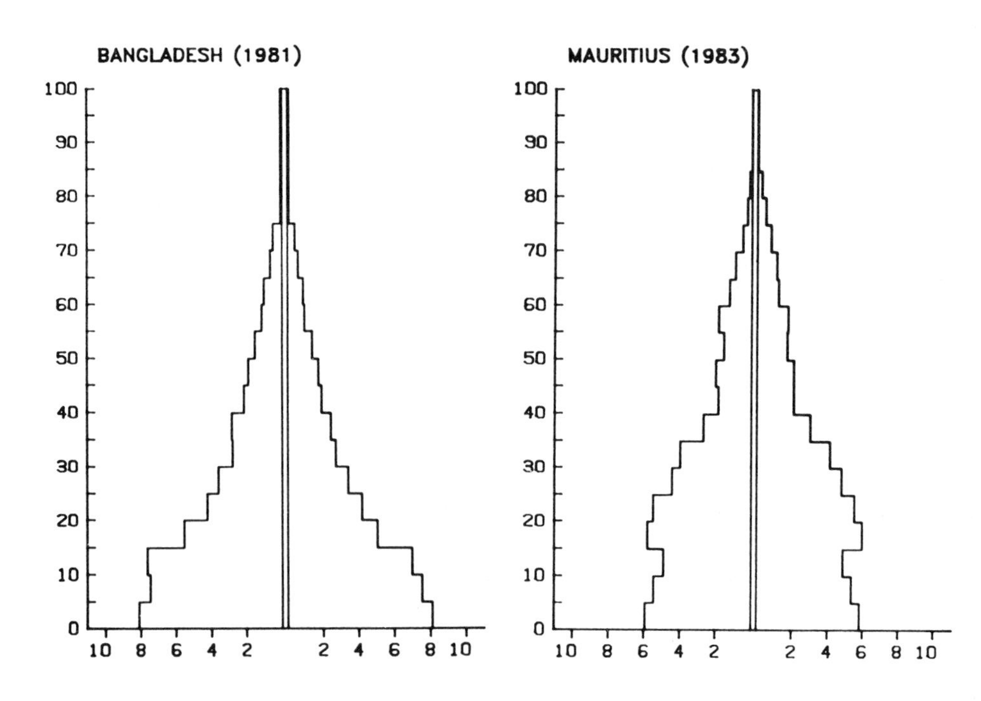
BANGLADESH (1981)
MAURITIUS (1983)
100
90
80
70
60
50
40
30
20
10
0
10 8 6 4 2 2 4 6 8 10

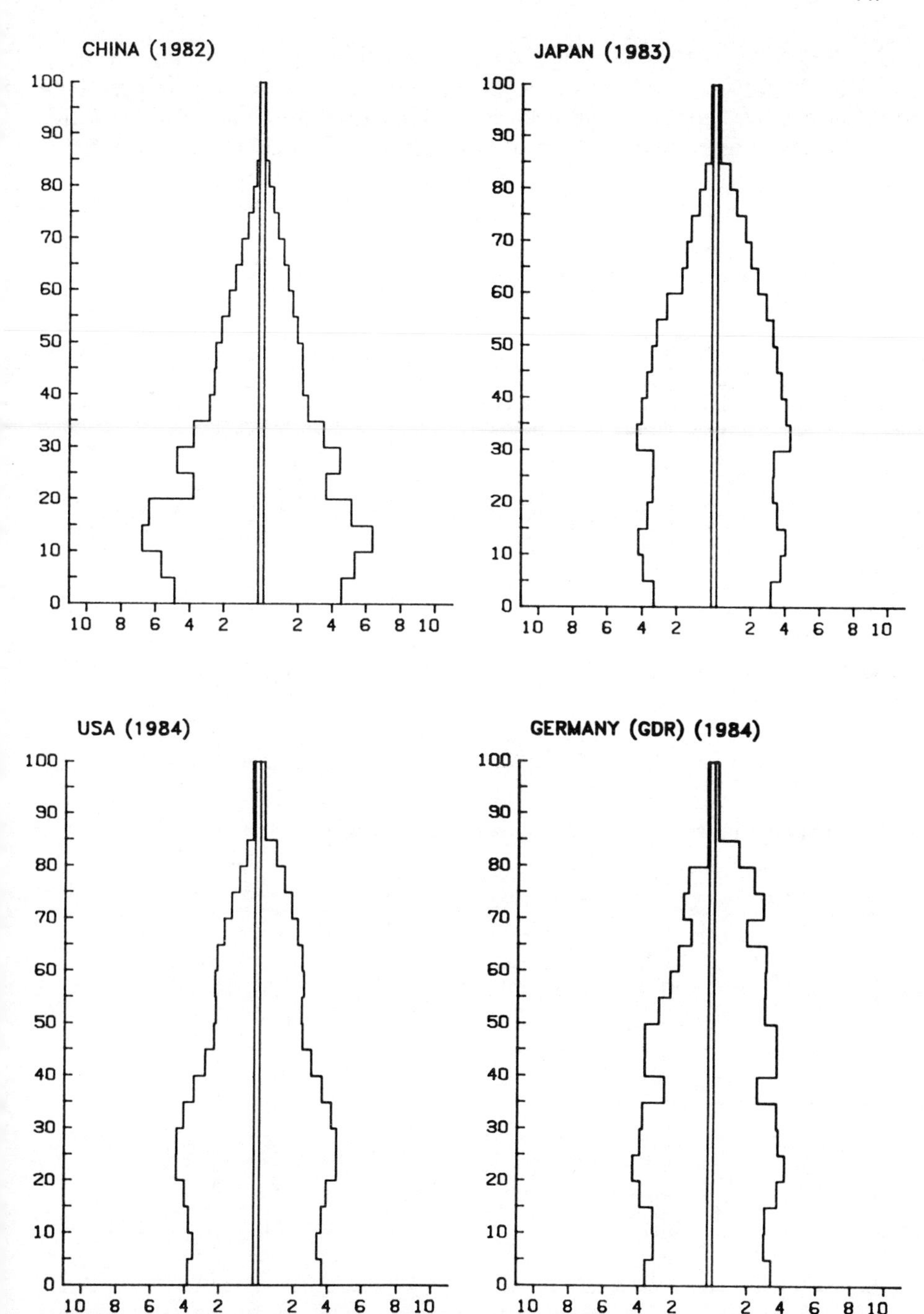

Fig. 9.2 Age–sex profiles for selected countries in the early 1980s. Horizontal axes represent percentage population; vertical axes show age in years

Fig. 9.1 show how these phenomena continue to show up in the profile. The post-1945 'baby bulge' is YY in Fig. 9.1(d) and the bulge of the early 1960s boom ('You've never had it so good' – Harold Macmillan) is shown at ZZ in Fig. 9.1(e).

Whereas Fig. 9.1 shows absolute numbers for the same country at different points of historical time, Fig. 9.2 gives *percentages* so that structural comparisons can be made among countries with very different total populations at roughly the

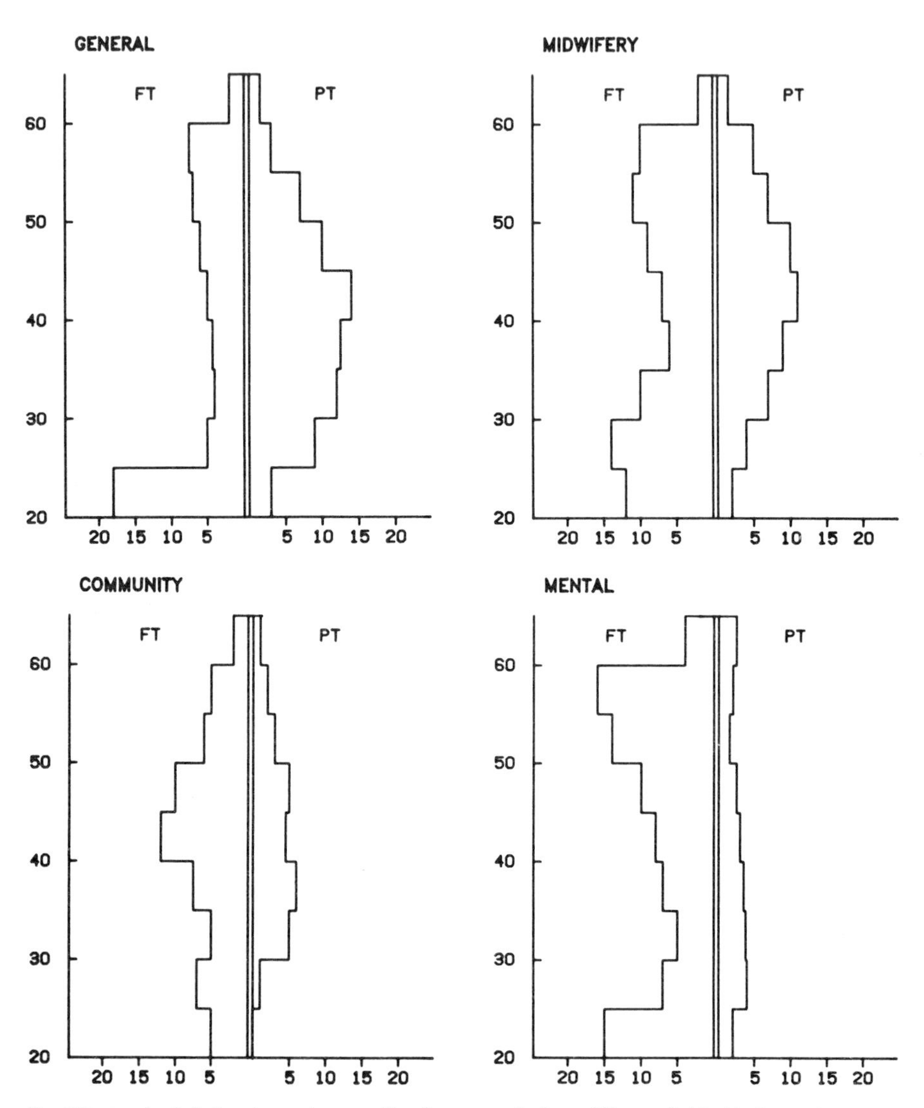

Fig. 9.3 Age by full-time/part-time profiles for nurses in four different fields. Horizontal axes are percentages; vertical axes show age in years

same time. The sequence demonstrates the range of shapes that are found simultaneously in countries at different stages of development. The Kenyan picture, with 51% of the population under the age of 15, is typical of a Third World country; the fact that the Mauritius pyramid of 25 years ago was very similar to that of Kenya today demonstrates the rapidity with which a population structure can change. In China the effect of the Cultural Revolution and a recent vigorous birth limitation policy can be seen, as can a massive post-war birth control programme by abortion in Japan to direct resources to industrial development. The effect of the Second World War on the male Japanese population is also just detectable. The USA picture shows a slightly distorted beehive shape due to recession followed by a post-war boom reflected in the birth rate. East Germany shows a sharp fall in the birth rate during the First World War and after the Second and the effect of the Second World War on the males of fighting age.

Simply to demonstrate the usefulness of these diagrams in another context, Fig. 9.3 shows age profiles for trained nurses in a health board area working in four different fields. Instead of sex, the division about the centre line is between full-time (FT) and part-time (PT). Notice how the numbers in general nursing (mostly female) swing towards part-time during child-rearing years. In midwifery, the delay due to a specialized training period and the slightly greater proportion of unmarried women modifies the pattern slightly. In the psychiatric field the higher proportion of male nurses and hours more conducive to full-time working for women present another pattern.

9.5 COHORTS

A 'snapshot' sequence of pyramids may reveal the effects of social (e.g. family planning), historical (e.g. wars, plagues, famines) and economic (e.g. recessions) changes on population structure, but it is also useful to think dynamically in terms of cohorts.

A group of people who entered some state at about the same time and whose life histories are thereafter studied is called a **cohort**. A cohort may consist of all persons born in the same year (a **birth cohort** or **generation**); but all women married in a particular year or a random sample of all cancer sufferers diagnosed in a certain month are equally valid examples of cohorts. Cohorts are used to relate events to those that must necessarily precede them, e.g. 'termination of a marriage' must precede 'remarriage'. Only in this way can we study the probability of the second event occurring. However, such a probability may change against any of three time scales:

1. the passage of *historical* time ('calendar' time)
2. the *age* of the individual ('maturing')
3. the *elapsed time* since the prior event.

The probabilities must therefore be expressed in terms of specific cohorts.

SUMMARY

The usefulness of demographic techniques in planning for human societies and large workforces or warning of impending problems in animal populations is considerable. Demographic change throughout phases of industrialization, etc., is of particular interest.

EXERCISES

1. Find recent figures giving the age–sex population breakdown for Australia, Canada or New Zealand and draw the population pyramid with the horizontal axis as a percentage. Compare the pattern with those given in Fig. 9.2 and comment on any special features.
2. Figure E9.1 gives the age-sex profiles for two countries that have experienced a good deal of immigration. Investigate the recent histories of these countries and try to explain the patterns.
3. Figure E9.2 shows the age–sex profiles of two island communities. Discuss possible explanations for any interesting features of the two pyramids.

(*Hint:* Remember that the entire population of Tristan da Cunha (how many?) was temporarily evacuated to the UK in 1961.)

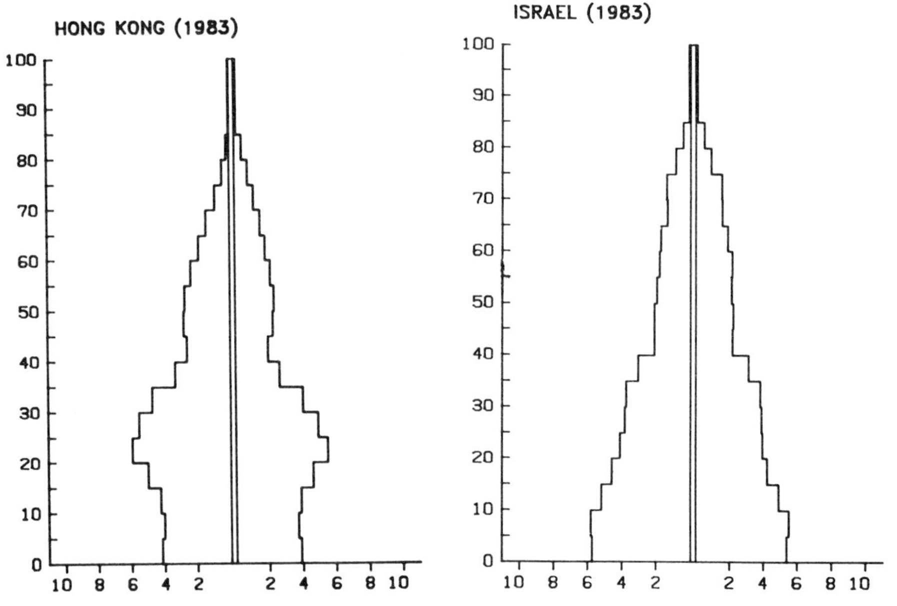

Fig. E9.1 Age–sex profiles for Hong Kong and Israel. Horizontal axes represent percentage population

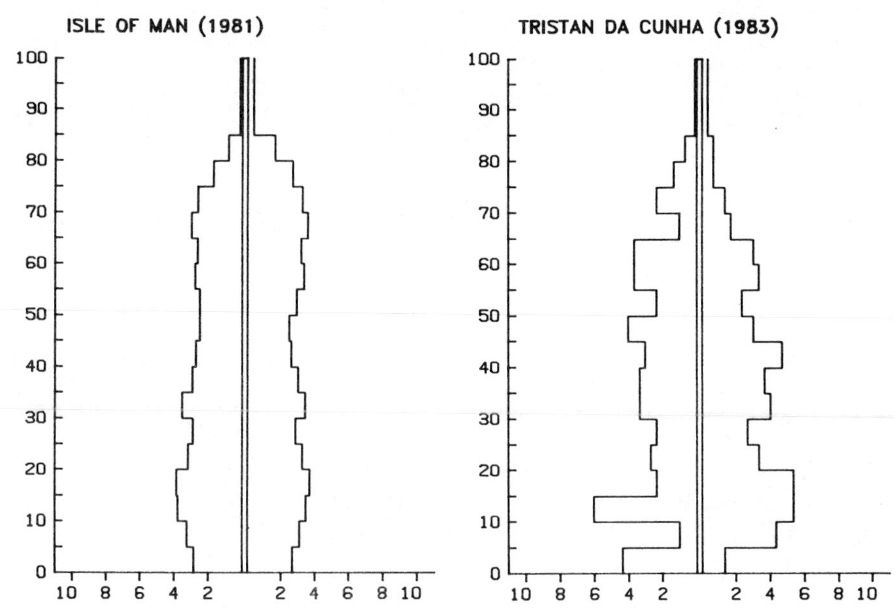

Fig. E9.2 Age–sex profiles for the Isle of Man and Tristan da Cunha. Horizontal axes represent percentage population

4. Table 2.7 of the *Employment Gazettes* of 1981 and 1986 give the following figures for unemployment (thousands) in the UK in April of 1979 and 1986:

Age	< 18	18–19	20–24	25–34	35–44	45–54	55–59	⩾ 60
April '79								
males	38.2	64.3	144.5	206.0	133.4	124.4	75.2	130.3
females	35.1	53.1	93.7	78.2	35.6	41.5	25.1	1.2
April '86								
males	107.1	185.2	438.9	548.8	384.1	323.4	226.4	76.2
females	79.5	129.4	243.7	256.4	126.0	124.3	74.6	1.0

Draw age–sex profiles for the two years. (Should you use absolute numbers or percentages on the horizontal axis?) Compare the patterns. Find out what you can about the definition of 'unemployment' in the two years.

10
Components of population change

10.1 DEATH

We shall begin our discussion of components of population change by looking at mortality since death is a well-defined event that happens to everyone. We shall initially assume a closed population, i.e. zero migration.

10.1.1 Crude death rates

A simple measure of the incidence of death during some period of time is given by the **crude death rate** (CDR):

$$\text{CDR} = \frac{\text{no. of deaths in period}}{\textit{average} \text{ population size in period}} ‰$$

The sign ‰ means that the rate is expressed per thousand. The period is almost always a calendar year; the average population size may be the estimated mid-year figure as in the UK or the average of the 1 January figures in the year in question and the following year.

Scottish figures for 1984 were as follows:

	Male	Female	Total
No. of deaths	30 731	31 614	62 345
Estimated mid-year population (thousands)	2 483.5	2 662.2	5 145.7

Source: Table A1.3 of 1984 *Annual Report of Registrar General for Scotland.*

$$\text{Overall CDR} = \frac{62\,345}{5\,145\,700} \times 1000 = 12.12‰$$

$$\text{Male CDR} = \frac{30\,731}{2\,483\,500} \times 1000 = 12.37‰$$

$$\text{Female CDR} = \frac{31\,614}{2\,662\,200} \times 1000 = 11.88‰$$

The last two rates illustrate *Male excess mortality*, a common phenomenon of most countries this century for all age groups except where death in childbirth reverses the pattern.

Here are some other CDRs (from Tables A1.2 and A2.4 of the 1984 *Annual Report of Registrar General for Scotland* and Table 18 of the *UN Demographic Yearbook*, 1984):

Scotland (1855–60 av.)	20.8
Scotland (1901–05 av.)	17.1
England & Wales (1984)	11.4
Isle of Man (1983)	14.8
France (1984)	9.8
Spain (1982)	7.4
USA (1983)	8.6
Costa Rica (1983)	3.9
Puerto Rico (1983)	6.6
Israel (1983)	6.8
Japan (1983)	6.2
USSR (1984)	10.8
Hong Kong (1984)	4.8
Gambia (1980–85 av.)	29.0

Why are Hong Kong and Costa Rica so low? Why is Spain lower than France, and Puerto Rico lower than USA?

10.1.2 Age-specific death rates

What is obscured by the CDR is the nature of the age structure of the population and the differential mortality involved. The risk of death is high for the first year of life, falls sharply to reach a minimum during the teens then climbs, especially from middle age onwards. Other factors are also involved; risks tend to be greater for males than for females at all ages (see Table 10.1); marital state, occupation and social class are likewise important.

Thus, populations with a high proportion of the very young and the elderly can expect (other things being equal) to have higher CDRs than populations largely composed of young and middle-aged people. We must therefore consider the rates associated with different age groups – **age-specific rates**.

Suppose two towns A and B, each with a population of 100 000 are as follows:

Town A

Age	Population (P_a)	Deaths	Age-specific rates (m_a)
< 5	20 000	100	5.0
5–45	50 000	50	1.0
> 45	30 000	900	30.0
Overall	100 000	1050	10.5 (CDR)

Town B

Age	Population (P_b)	Deaths	Age-specific rates (m_b)
< 5	25 000	100	4.0
5–45	40 000	20	0.5
> 45	35 000	980	28.0
Overall	100 000	1100	11.0 (CDR)

Notice that in each age group, town B has lower specific mortality than A yet the CDR is higher because of the difference in age structure.

Age-specific rates are often calculated as averages over three years, partly to reduce instability of the statistics in older, almost empty, age classes, and partly to obviate problems due to especially bad weather at the end of a year causing early demise of those who would otherwise die the following year.

10.1.3 Standardized death rates

Age-specific mortality rates (usually in 5-year age groups) are interesting in themselves but cumbersome as a means of summarizing death rate patterns. How can we take into account the age (and sex) structure of a population in a single figure?

Notice that the CDR can be thought of as a weighted mean of the age-specific

Table 10.1 Age-specific death rates for males and females separately; Scotland 1911–15 (average) and 1984: rates per 1000 population

	Male		Female	
Age	1911–15	1984	1911–15	1984
0–	140.3	11.9	111.5	9.1
1–4	19.0	0.6	18.0	0.5
5–9	3.8	0.3	3.8	0.2
10–14	2.5	0.3	2.6	0.2
15–24	4.2	0.8	3.9	0.3
25–34	5.7	1.2	5.6	0.6
35–44	8.6	2.2	8.1	1.5
45–54	15.7	7.2	12.7	4.4
55–64	31.4	20.6	25.0	12.2
65–74	68.3	52.0	54.6	28.3
75–84	142.2	110.6	120.4	68.4
85+	280.5	221.5	262.1	177.8
All ages	16.2	12.4	15.2	11.9

Sources: *Registrar General for Scotland Annual Reports* 1973 and 1984, Table B1.1.

rates, using age class population numbers as weights, i.e.

$$\text{CDR} = \frac{\sum(P_a \times m_a)}{\sum P_a}$$

When we want to compare two populations, as with the example of towns A and B above, we must apply the same (standard) weighting scheme to both; this gives a **standardized death rate** (SDR). Which standard?

1. We could use the age structure of one of the towns (say town A) for both rates. Thus, the rate for B would be

$$\frac{\sum(P_a \times m_b)}{\sum P_a}$$

i.e. $$\text{SDR} = \frac{20\,000 \times 4.0 + 50\,000 \times 0.5 + 30\,000 \times 28.0}{100\,000}$$

$$= 9.5$$

which is now consistent with the fact of uniformly lower age-specific rates than A. You can check that standardizing A by B's population gives a rate for A of 12.2 against B's 11.0.

2. We could combine the populations and standardize both rates using this (fictitious) age structure. Thus, for A we have $\sum(P_a + P_b) \times m_a / \sum(P_a + P_b)$, i.e.

$$\text{SDR} = \frac{(20\,000 + 25\,000) \times 5.0 + (50\,000 + 40\,000) \times 1.0 + (30\,000 + 35\,000) \times 30.0}{200\,000}$$

$$= 11.3$$

and, for B, the rate is 10.2.

3. For both towns, we could use some weighting scheme derived externally, e.g. the population structure for the whole country. For international comparability, a standard population fixed by the International Statistical Institute is usually used.

10.1.4 Standardized mortality ratios

It is now only a short step to think of comparing 'local' specific mortality patterns (m_a) with a standard (m_s) by means of a weighted average over the age groups of the ratios of m_a to m_s, using as weights the deaths that would be expected locally if the standard specific mortalities were to apply, i.e. $m_s \times P_a$. This is called the **standardized mortality ratio** (SMR). That is,

$$\text{SMR} = \frac{\sum(m_a/m_s) \times m_s P_a}{\sum m_s P_a} \times 1000 \qquad \text{(NB Conventional multiplier is 1000)}$$

$$= \frac{\sum m_a P_a}{\sum m_s P_a} \times 1000$$

$$= \frac{\text{Actual local deaths}}{\text{Expected local deaths if standard mortalities applied}} \times 1000$$

This summary figure is quite widely used by Registrars General in the United Kingdom but others are possible. Notice how the SMR is analagous with the Paasche index of Section 8.3.2.

10.1.5 Cohort rates and probabilities

Death rates can be criticized in that the denominator population is a somewhat fluid concept. Suppose we want a death rate specific to 40-year-old males in 1986; then we really need to know, not only how many men aged 40 completed years died in 1986, but the average number of 40-year-old men alive in 1986. Even if this number could be regarded as reasonably constant throughout the year, it is clear that the group of individuals exposed to the risk of dying is constantly changing. A man who became 40 just before or just after 1 January 1986 is much more likely to contribute a death to the numerator than a man who became 40 in the early days of 1985 or the last days of 1986.

Such a problem is less serious when 5-year age groups are considered (since entries and exits form a smaller proportion of the denominator population) and even less serious for the whole-life CDR; but it is at its worst for children in their first year of life. (In an extreme case we might have a fairly constant number of 0-year-old children (say 5000) in some population, yet averaging 100 deaths each week in the 0–1 age group, so that the ratio of deaths to mean population over a year exceeds 1.01).

Consider as an example the following data:

No. of men aged 40 on 1 January 1986 = 298 000
No. of men aged 40 on 1 January 1987 = 294 000
No. of male deaths at age 40 in 1986 = 1160

Then the 'average' male population aged 40 in 1986 can be estimated as

$$\frac{(298\,000 + 294\,000)}{2} = 296\,000$$

Hence the male crude death rate at age 40 in 1986 is

$$m_{40} = \frac{1160}{296\,000} \times 1000 = 3.92‰$$

What is generally of more interest is the *probability*, q_{40}, of death between ages 40 and 41 for men in the 1946 birth cohort. From the available information, we cannot tell how many of the 1160 deaths in 1986 belong to the 1946 cohort so we can only estimate it to be 1160/2 = 580 and assume that another 580 of the 1946 cohort will die in 1987. We must assume also that the number of the 1946 cohort alive on 1 January 1987 is 296 000 so that the total number of the cohort who achieved their 40th birthday was 296 000 + 580. Thus

$$q_{40} = \frac{1160}{296\,580} = 0.003\,91$$

It can be shown algebraically (and these calculations verify) that

$$q = \frac{2m}{2000 + m} \qquad \text{(where } m \text{ is the rate per thousand)}$$

so that cohort probabilities can be estimated from mortality ratios.

10.1.6 Life tables

Once the cohort probabilities have been established, a **cohort life table** can be constructed; a birth cohort (or **radix**) of arbitrary size, say 10 000, is depleted at successive ages according to the cohort death probabilities, q. Table 10.2 shows such a display for a robin population. (The robin has a conveniently short lifespan.)

The columns headed q_x, l_x, and d_x present no problem; given the information in any one of them, the other two can be easily found.

Table 10.2 Life table for robins

Age x	Prob. of death q_x	Survivors to age x l_x	Deaths d_x	Expectation of life at age x, e_x	T_x
0	0.7	10 000	7000	1.1558	11 558
		(−7 000)			
1	0.5	= 3 000	1500	1.686	5 058
		(−1 500)			
2	0.3	= 1 500	450	1.872	2 808
3	0.4	1 050	420	1.46	1 533
4	0.5	630	315	1.1	693
5	0.8	315	252	0.7	220.5
6	1.0	63	63	0.5	31.5

The column headed e_x requires some explanation; it represents the average length of life remaining to a bird that has attained its xth birthday. Thus, e_0 represents the life expectancy at birth.

The T_x column has been included merely to help explain the calculation of e_x; it represents the total 'robin-years' of life remaining for all of the original cohort that have reached age x. Let us assume that deaths occur at completely random points throughout the year. This is clearly most suspect at the extremes of the age range. However, if it is accepted, then a bird that dies at age x (in completed years) lives on average $(x + 1/2)$ years. Hence, for the last line of a life table, e_x must always be 0.5. The 63 robins that die at random points in their seventh year of life must therefore live a total of $T_6 = 63 \times 0.5 = 31.5$ robin-years. Now consider T_5; the 252 robins that die aged 5 live a total of 252/2 robin-years, to which must be added one year of life for each of the 63 birds that survive at least another year plus their total life beyond their next birthday (i.e. T_6).

$$\therefore\ T_5 = 252/2 + 63 + 31.5 = 220.5$$

In general,

$$T_x = d_x/2 + l_{x+1} + T_{x+1}$$

thus allowing completion of the T_x column from the bottom upwards.

Thereafter, e_x is easily found as T_x/l_x. Remember that e_x is the average *remaining* life conditional on a robin reaching age x. It is therefore quite possible (as here) to have

$$e_x < e_{x+1}$$

Cohort life tables are little used for humans because:

1. it takes too long to follow a human birth cohort through life – the information is of little interest at the end of such an exercise

2. mortality depends less on birth cohort and more on year-to-year climatic, economic, epidemiological, etc., variations.

Hence, it is usual to calculate the q's for the current year or a period of years and apply these to what becomes a hypothetical cohort 'followed' through a **current** or **period life table**. In such a case, notice that the average length of life, e_0, in a population is not, in general, equal to the average age at death because all the birth cohorts have been compounded irrespective of their sizes.

10.1.7 Infant mortality

Special problems are associated with calculations of infant mortality because of the rapid decline in the death rate during the first year of life. We shall not discuss such special calculations.

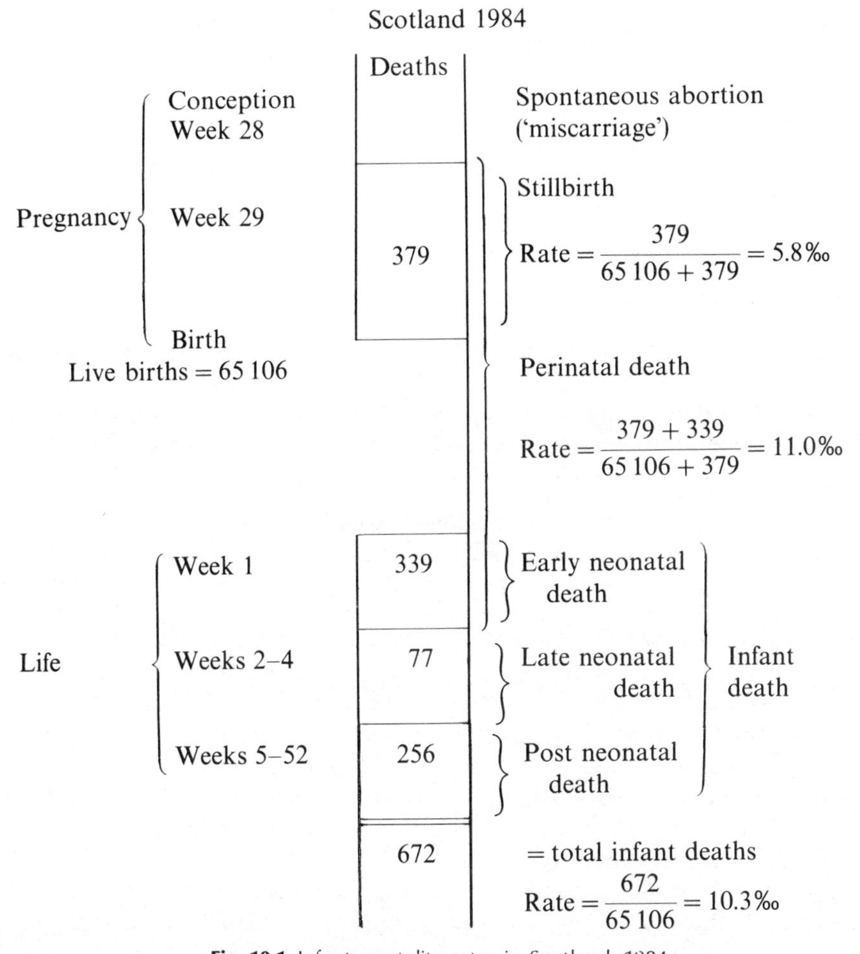

Fig. 10.1 Infant mortality rates in Scotland, 1984.

The associated jargon and rates may be introduced by looking at the following figures for 1984 extracted from Tables D1.1 and D2.1 of the 1984 *Annual Report of the Registrar General for Scotland.* (Stillbirth and perinatal death rates are per thousand births, live or still; the infant death rate is per thousand live births.)

Table 10.3 shows a selection of infant death rates.

Table 10.3 Some infant death rates (‰)

Scotland (1984) males	11.7
Scotland (1984) females	8.9
England & Wales (1983)	10.1
Netherlands (1983)	8.4
Yugoslavia (1981)	30.8
USA (1980)	12.6
Pakistan (1979)	94.5
Mali (1976)	120.9

Sources: 1984 *Annual Report of Registrar General for Scotland* and *UN Demographic Yearbook*, 1984.

10.2 MARRIAGE

Many of the summary techniques used to describe mortality carry over into the other components of change. Strictly speaking, marriage does not contribute directly to population change, but, because of its association with the legitimacy of births in most countries, marriage rates have been of some interest. The crude marriage rate is thus

$$\text{CMR} = \frac{\text{No. of persons married in period}}{\text{Average population}}$$

(The rate of actual marriages is half this, but the definition is more consistent with the CDR. For the same reason, one does not normally restrict the denominator population to those who are of appropriate age and single.)

Naturally, age/sex specific rates and probabilities can be calculated. In finding such probabilities, some account should be taken of deaths in the cohort during the year in question; we can assume that a person dying lives, on average, six months of the year so that the population at risk of marriage at the start of the year should be reduced by half the deaths. This correction is usually negligible for the most marriageable age groups.

Marriage tables (analagous to life tables) can be constructed showing, as might be expected, that advancing age first increases the chance of an imminent marriage but is finally an unfavourable factor. However, in contradistinction to mortality summaries, period tables have less to commend them in the case of marriage where year-to-year variations may be considerable and past history affects cohorts differentially. For example, if there is an imbalance of the sexes in an age group either through natural variation in the sex ratio or because men have been killed during war or are at war, the overall marriage rate for that age will tend to fall temporarily and that for the less numerous sex will rise.

In the construction of a marriage table, notice that not all of the radix cohort will marry and that a marriage table can stop at about 55 years (certainly for first marriages) since the marriage probability has become negligible by that age.

10.3 FERTILITY (live births)

Birth statistics can be expressed by a number of different measures.

10.3.1 Crude birth rate (CBR)

Unless there is a very high proportion of females of child-bearing age, the CBR is usually less than 50‰. Table 10.4 gives some typical figures. Remember that 'natural increase' = CBR − CDR.

10.3.2 General fertility rate

This is defined as

$$\frac{\text{No. of live births}}{\text{No. of women aged 15–49}}$$

(Sometimes 15–44 years; sometimes married and non-married separately.) The figure was 51.5‰ in 1984 in Scotland.

10.3.3 Age-specific fertility rates

These are as for mortality figures. Standardized rates and 'standardized fertility ratios' are sometimes calculated. Thus, for Scottish legitimate births:

Age group	16–19	20–24	25–29	30–34	35–39	40–44
Year 1939	525.7	336.0	220.0	140.5	82.7	25.8
1984	299.5	226.4	167.0	75.3	21.5	3.7

Note, however, that in the presence of birth control, fertility is almost more dependent on duration of marriage than on age itself.

Table 10.4 Selected crude birth rates (‰)
(a) Historical figures for UK

	CBR	CDR	Natural increase = CBR − CDR
1870–2	35.0	23.3	11.7
1932	15.8	12.9	2.9
1939	15.3	13.1	2.2
1940	14.6	16.2	−1.6
1947	20.7	13.6	7.1
1960	17.5	12.1	5.4

Source: *Annual Abstract of Statistics*, 1946, 1971

(b) International comparisons in 1983

	CBR	CDR	Natural increase
England & Wales	12.7	11.7	1.0
Scotland	12.6	12.3	0.3
Northern Ireland	17.3	10.2	7.1
Eire	19.0	9.8	9.2
Sweden	11.0	10.9	0.1
Albania	26.0	6.1	19.9
East Germany	9.7	11.7	−2.0
USSR	20.0	10.4	9.6
USA	15.4	8.6	6.8
Bolivia	44.0	15.9	28.1
Costa Rica	30.0	3.9	26.1
Lesotho	55.1	16.4	38.7
Mainland China	18.5	6.8	11.7
Bangladesh	44.8	17.5	27.3

Source: *UN Demographic Yearbook*, 1984.

10.3.4 Total fertility rate

This is simply the unweighted sum of the age-specific rates. It assumes each age to be of equal significance and represents the total number of children born (on average) to each woman of a hypothetical cohort throughout her life. If specific rates refer to 5-year age groups, we must assume that the rate applies in every year of the group so that the rate derived from the fictitious figures of Table 10.5 is

$$5 \times 0.58 = 2.9 \text{ (children)}$$

Table 10.5 Births in some population

Age group	Number of women	Number of live births	Age-specific birth rate	Probability of survival to start of age group
15–19	10000	200	0.02	0.93
20–24	10500	2100	0.20	0.91
25–29	9750	1755	0.18	0.90
30–34	9500	950	0.10	0.89
35–39	9200	552	0.06	0.87
40–44	9150	183	0.02	0.84
			$\sum = 0.58$	

10.3.5 Gross and net reproduction rates

Some of these 2.9 children will be male and some female; if we assume a birth sex ratio of, say, 52:48, i.e. probability of female birth = 0.48, then the expected female births per woman is

$$2.9 \times 0.48 = 1.392 \text{ (daughters)}$$

This **gross reproduction rate** measures the population's capacity to replace itself assuming fertility does not alter.

However, this makes no allowance for deaths of the mothers or deaths before 15 among the daughters replacing them. The 'survival' column of Table 10.5 based on current mortality experience, at the time of the calculation, can be used to reduce the fertility figures and give the **net reproduction rate**

$$= 5 \times 0.48 \times (0.93 \times 0.02 + 0.91 \times 0.20 + \cdots + 0.84 \times 0.02)$$
$$= 5 \times 0.48 \times 0.5206 = 1.249 \text{ (daughters)}$$

10.4 MIGRATION

Migration is usually the least important component of population change. The movement may be international (e.g. UK to USA) or inter-regional (e.g. New South Wales to Western Australia) or 'local' (e.g. North London to South London). In all these cases the basic information comes from a census question such as 'Where were you living five years ago?' But what happens to the five years immediately following a decennial census? Likewise, some moves during the five-year pre-census period will be missed. In any event, memory lapses are likely to present major reliability problems. In the UK, the International Passenger Survey (see Section 6.1.3) provides some information but only on international movements. Local changes can be detected from employment, taxation and social security records and also Executive Councils' lists of general practitioners' patients (though these are notoriously out of date especially for age groups having little need of medical care and therefore slow to register with a new GP after moving).

The causes of migration are generally of interest. There can be economic pressures such as industrialization that cause rural to urban movement. Career advancement (e.g. promotion to head office) causes drift to the capital city but can also cause urban to suburban movement. In this country, age-related factors such as ill health and retirement may initiate a move to climatically favoured areas such as the south-coast resorts. Military service also brings about major and frequent changes of domicile. International arrangements may be made to cater for refugees such as the Vietnamese 'boat people'; the insecurity of religious minorities (e.g. the Sikhs in India) also causes redistribution of population.

Net migration (i.e. $I - E$) for the period between censuses is easily obtained by subtraction of the natural increase ($B - D$) from the change in total population but will be subject to considerable error in comparison with the other components of change. Even if in-migration balances out-migration so that ($I - E$) is zero, the differences in life tables, fertility, etc., of the incoming and outgoing migrants may show up. This phenomenon is simply but unhelpfully called **the demographic effect**.

SUMMARY

The four components of population change are birth, death, immigration and emigration. The first two of these are generally the most important and give rise to rates and methods for temporal and international comparison. Cohort death rates and probabilities are needed for the construction of life tables which are a vital part of actuarial and insurance work. In developed countries, the birth rate fluctuates much more than the death rate (due to social and economic changes) and is the major cause of difficulty in population projection.

EXERCISES

1. Calculate the standardized mortality ratio for barmen aged 25–64 for the quinquennium 1949–53 from the following data:

Age group	Number of deaths of barmen, England & Wales 1949–53	Census population of barmen (1951)	Number of deaths males, England & Wales, 1949–53 (thousands)	Census population males (1951) (thousands)
25–34	38	2790	25	3140
35–44	75	2900	47	3291
45–54	179	2580	118	2874
55–64	254	1620	233	2028

2. The populations of two towns A and B and the number of persons who die in them in a certain year are given in the table below. Describe the populations and their deaths and compare their mortality experience using S as the standard population. Discuss the result.

	A		B		S	
Age	Popln.	Deaths	Popln.	Deaths	Popln.	Deaths/1000
0–24	15 000	25	10 000	20	12 000	1.5
25–49	12 000	60	8 000	50	11 000	4.0
50–74	9 000	360	10 000	400	8 000	30.0
75 +	3 000	300	5 000	400	2 000	100.0
	39 000	745	33 000	870	33 000	

3. The information in Table E10.1, from a publication of the Office of Population Censuses and Surveys, gives all teenage conceptions leading to maternities or abortions (under the 1967 Act) in England and Wales in 1969 and 1983. The conception rates are given per 1000 women at each age (age 13–15 for 'under 16'). Numbers are in thousands.

Table E10.1 Teenage conceptions resulting in maternities and abortions

		Total conceptions		Maternities		Abortions	
Year	Age at conception	Number	Rate	Number	Rate	Number	Rate
1969	< 16	6.5	6.8	4.9	5.1	1.7	1.7
	16	14.3	44.1	12.4	38.4	1.8	5.6
	17	24.5	76.8	22.3	69.9	2.2	6.9
	18	34.6	103.9	31.9	95.8	2.7	8.0
	19	43.6	125.5	40.7	117.3	2.9	8.2
1983	< 16	9.4	8.3	4.0	3.6	5.3	4.7
	16	15.2	38.7	8.4	21.4	6.8	17.3
	17	22.6	56.0	14.5	35.9	8.1	20.2
	18	29.9	72.2	21.2	51.2	8.7	20.9
	19	35.3	84.8	26.9	64.5	8.4	20.3

(a) Make *three* different estimates of the number of 16-year-old women in 1969. Why do they differ and which is best?

(b) Construct indices to compare overall teenage abortion in 1969 and 1983
 (i) per thousand girls
 (ii) per thousand conceptions.
 What does each tell you?

4. Members of a random sample of 1000 men aged 65 in 1955 were followed up and the numbers still alive after successive 5-year intervals were found to be as follows:

1960	800
1965	520
1970	260
1975	52
1980	3
1985	0

Construct a life table for these men including estimates of life expectancy at the start of successive 5-year age periods. Are your estimates all equally unreliable?

Discuss the disadvantages of using these figures to help determine the provision of old people's facilities in your local community over the next 20 years.

How would you seek better information?

5. In the little-known country of Academia, it is a well established fact that 1 in 10 undergraduates leave university during or at the end of their first year. The corresponding figure for second-year undergraduates is 1 in 9. By the end of the third year only 50% of the original starters have not left. No undergraduate may stay longer than four years.

Draw up a life table for a cohort of 5000 students showing, for each year of study, the probability of leaving and the expected remaining stay calculated on the assumption that one-tenth of the leavers in any year depart at random times during the year and the remainder go at the end of that year.

6. An article in the September 1987 issue of *Employment Gazette* gives the following probabilities of reaching particular durations of unemployment based on benefit claims in 1985.

Duration (months)	Age	
	< 25 years	≥ 25 years
3	0.42	0.59
6	0.26	0.42
9	0.17	0.32
12	0.12	0.26
15	0.07	0.17
18	0.06	0.14
24	0.03	0.10
36	0.01	0.07

Construct and compare the two 'life tables' for length of unemployment. What are you assuming?

7. For Table E10.2 of death rates (*OPCS Monitor* DH1 80/3).

(a) explain the terms 'ICD 180' and 'birth cohort',

Table E10.2 Cancer of the cervix uteri (ICD 180), females. Table of deaths per million population with birth cohorts indicated on the diagonals

	Age group																		
Year of death	0-	5-	10-	15-	20-	25-	30-	35-	40-	45-	50-	55-	60-	65-	70-	75-	80-	85 and over	
1951–55	0	—	—	0	1	10	30	58	93	136	203	254	285	304	315	361	327	315	
1956–60	0	—	0	0	1	9	37	74	119	154	181	197	246	284	313	336	366	361	
1961–65	0	—	—	—	1	5	18	67	134	180	187	178	222	232	274	301	332	357	1871
1966–70	0	—	—	0	2	7	15	44	104	176	204	201	193	217	247	271	290	351	1876
1971–75	—	—	—	—	3	10	22	38	67	130	190	199	199	193	206	247	257	262	1881
1976–78	—	—	—	0	2	14	30	51	59	91	175	198	217	198	209	220	249	247	1886
Birth cohort (diagonal)						1956	1951	1946	1941	1936	1931	1926	1921	1916	1911	1906	1901	1896	1891

(b) draw a graph to show how death rate changed with year of birth for the 30–34 and 40–44 year age groups,
(c) draw a graph to show changes in death rate with age at death for the 1901 and 1921 birth cohorts,
(d) what do these graphs tell you? What else can you infer from the table?

11
Population projection

11.1 RELIABILITY OF PROJECTIONS

Here is an early demographer, Gregory King, writing in 1692 on the future growth of the population of England.

> That, when the Romans invaded England, 53 years, before our Saviour's time, the kingdom had about 360,000 people; and, at our Saviour's birth, about 400,000 people;
>
> That, at the Norman Conquest, Anno Christi 1066, the kingdom had somewhat above two millions of people;
>
> That, Anno 1260, or about 200 years after the Norman Conquest, the kingdom had 2,750,000 people, or half the present number; so that the people of England have doubled in about 435 years last past;
>
> That in probability the next doubling of the people of England will be in about 600 years to come, or by the year of our Lord 2300; at which time it will have eleven millions of people; but, that the next doubling after that, will not be (in all probability) in less than 12 or 1300 years more, or by the year of our Lord 3500 or 3600; at which time the kingdom will have 22 millions of souls, or four times its present number, in case the world should last so long,

It is very easy to laugh at such calculations; even King's pre-1692 figures were just wild guesses. Yet, to this day, the estimation of future population size is a hazardous exercise (though an essential one for local and national governments and indeed for the world as a whole).

A projection can be made only on the assumption that current levels or trends continue. King lived in an age when growth rates changed only slowly so that his greatest concern was the possible end of the world! But, today, rapid changes in life expectancy, provision of international aid and the availability of birth control make forecasting for periods greater than five years difficult. In the longer term, estimates are usually made to cover reasonable extremes of change by presenting three sets of projections – high, central and low. Of course, publication of projections is often sufficient in itself to *cause* changes, e.g. by alteration of government policy.

Basically, a starting population of known (census) size is defined in terms of its age, area of residence, etc., and future developments in that population are estimated by consideration of the components of change in the light of current experience. The greatest error is likely to derive from the migration component. Thus, the current life table probabilities of death, q_x, are applied to the age groups of the recurrent population plus a term for net inwards migration, i.e.

$$\text{Pop(age } x+1, \text{ year } n+1) = \text{Pop(age } x, \text{ year } n) \times (1 - q_x) + \text{net migrants aged } x$$

Migrant mortality is ignored. Births have to be estimated from current birth rates but related to marriage trends.

11.2 AN EXAMPLE

Consider these principles as applied to a population of ants. This will suffice to demonstrate possible approaches. Assume constant death rate of 250‰ at all ages except that all ants die before age 6.

Projection I

We are told CBR = 330‰ and given the starting population in 1987, the bold figures are projected.

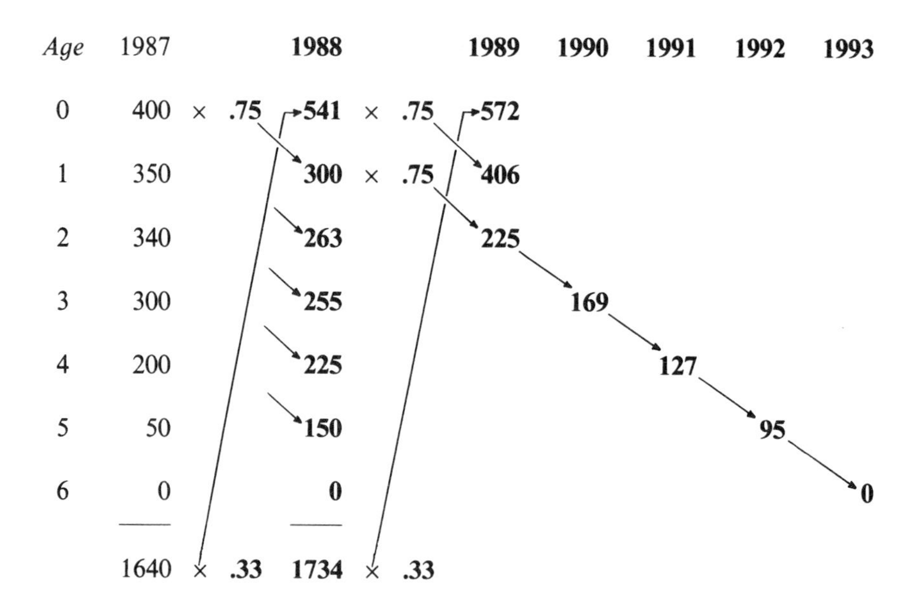

Projection II

Now assume that age-specific birth rates are given as age 1 = 850‰, age 2 = 715‰, zero otherwise. Then we have:

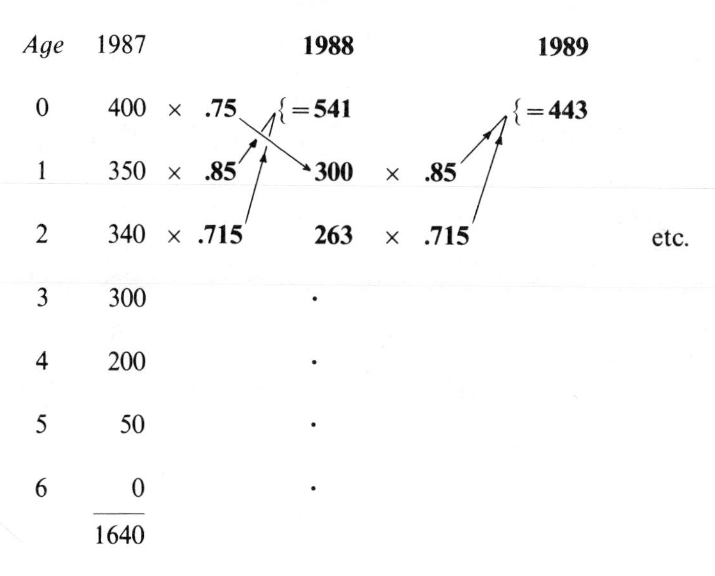

Projection III

Assume births = deaths, i.e. constant population size of 1640. This may seem a ridiculous assumption but, where there is an unvarying food supply, nature's response is often somewhat like this. Furthermore, if we were thinking about projecting the workforce of a large employer, the assumption corresponds to recruitment to replace leavers.

Age	1987	**1988**	**1989**
0	400	**447**	
1	350	**300**	**335**
2	340	**263**	**225**
3	300	**255**	etc.
4	200	**225**	
5	50	**150**	
6	0	**0**	
	1640	= **1640**	

so must have **447** births

Projection IV

Add to projection II in-migration of 100 ants/year during 1987, 1988 and 1989 having a fixed age structure

Age	
0	20
1	40
2	40

There is to be no fertility in year of arrival then fertility as for remainder of population. (Human populations can behave in this way.) Mortality as before.

Add to projection II:

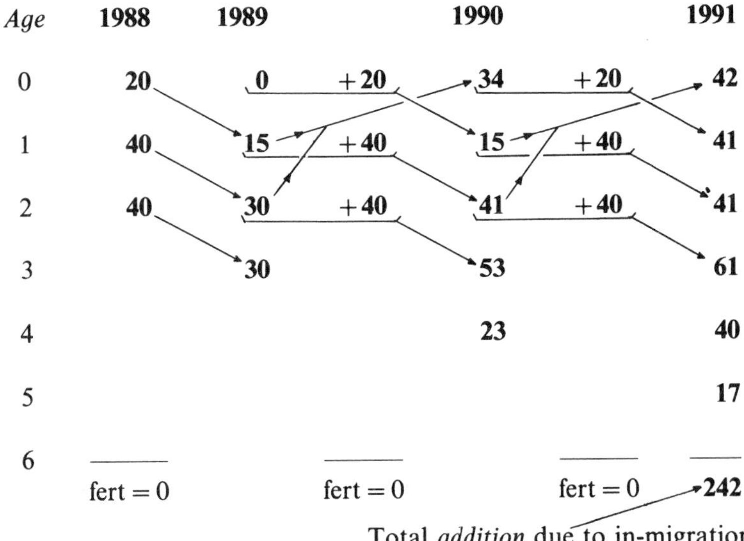

SUMMARY

Population projection is essential for planning but subject to considerable error due to the probable invalidity of assumptions about current and future trends. Methods are often somewhat *ad hoc.*

EXERCISES

1. In October 1987, a new university is opened in Academia (see Exercise 5 of Chapter 10) with 4000 first-year students. A suggested admissions policy for the next four years is that new first year students should be admitted each October to bring the *total* student numbers to 10% more than the previous October. Thereafter the total is to be held constant. Find the frequency distribution over years of study in 1991. Does the pattern seem satisfactory in the longer term?

2. Table E11.1 shows annual changes over the period 1977–85 in the number of live births, the crude birth rate, the general fertility rate, and the total period fertility rate; seasonally adjusted quarterly rates for 1984 and 1985 are also included.

Table E11.1 Live births in England and Wales, 1977–1985

Live birth occurrences								England and Wales	
Year and quarter		Live births		Crude birth rate		General fertility rate		Total period fertility rate	
		Number (OOOs)	% change from preceding year	Rate	% change from preceding year	Rate†	% change from preceding year	Rate*	% change from preceding year
1977		569.3	−2.6	11.5	−2.5	58.1	−3.7	1.66	−3.0
1978		596.4	4.8	12.1	4.8	60.1	3.3	1.73	4.2
1979		638.0	7.0	12.9	6.8	63.3	5.4	1.84	6.4
1980		656.2	2.9	13.2	2.7	64.2	1.4	1.88	2.0
1981		634.5	−3.3	12.8	−3.4	61.3	−4.5	1.80	−4.1
1982		625.9	−1.3	12.6	−1.3	59.9	−2.2	1.76	−2.0
1983		629.1	0.5	12.7	0.4	59.7	−0.4	1.76	−0.4
1984		636.8	1.2	12.8	1.0	59.8	0.2	1.75	−0.1
1985		658	3.4	13.2	3.3	61.4	2.6	1.79	2.2
1984	March	153.5	0.7	12.5	−0.5	58.7	−1.3	1.73	−1.5
	June	157.7	−2.2	12.5	−2.3	58.7	−3.1	1.72	−3.3
	Sept	167.1	2.3	13.0	2.2	60.5	1.4	1.77	1.1
	Dec	158.4	4.3	13.1	4.0	60.9	3.2	1.79	2.9
1985	March	160.4	4.5	13.2	5.5	61.5	4.7	1.80	4.4
	June	165.3	4.8	13.1	4.8	61.0	4.0	1.79	3.6
	Sept‡	173	3.4	13.4	3.2	62.0	2.5	1.81	2.1
	Dec‡	160	0.9	13.2	0.8	61.0	0.1	1.78	−0.2

Notes: All quarterly rates are seasonally adjusted.
†Births per 1000 women aged 15–44.
*The total period fertility rate is the average number of children which would be born per woman if women experienced the age-specific fertility rates of the period in question throughout their childbearing life span.
‡Estimated from births registered in the period.
Source: *OPCS Monitor* FMI 86/1

(a) Why do the live birth figures for the last two quarters of 1985 and for the whole of 1985 appear as rounded thousands? Why are the corresponding rates not similarly rounded?

(b) Write a report on the information in the table considering, in particular, future implications of the figures.

12
Changes with time – time series

12.1 INTRODUCTION

We are frequently presented with data consisting of observations at successive points of time; such data are said to form a **time series.** We shall consider here only time series having equal intervals between observations (e.g. numbers of road accidents in a city during successive hours of the day, daily climatic measurements, monthly export totals, quarterly unemployment figures, yearly coal consumption, etc.).

Data of this form usually have the (normally unwelcome) property that each observation is not independent of those that have preceded it. However, it may be possible to use this dependence to form the basis for short-term predictions of future behaviour.

12.2 SMOOTHING TIME SERIES

Time series data can be represented graphically in a **line diagram** of observations plotted against time as horizontal axis, successive points being joined by straight lines. For such diagrams, the time axis is often of considerable length and rules about the size relationship between horizontal and vertical axes can be relaxed.

Often the only striking feature of such a representation is the very erratic behaviour of the observations in relation to time.

Example 12.1

Consider the data on imported softwood given in Table 12.1. We require to clarify any inherent pattern that might be present in the variation of imports over time. We can do this by various systems of **moving averages**, whereby an observation is replaced by the mean of a number of observations centred on the one in question. For example, if we take 5-year moving averages, the 1956 figure is replaced by the mean volume of imports for years 1954 to 1958, i.e.

$$\frac{6902 + 7228 + 6829 + 6833 + 6421}{5} = 6842.6$$

Table 12.1 UK imports of softwood, 1954–84 (in thousands of cubic metres)

Year	Softwood	Year	Softwood	Year	Softwood	Year	Softwood
1954	6902	1962	7568	1970	8221	1978	6709
1955	7228	1963	7804	1971	8181	1979	7053
1956	6829	1964	9103	1972	8630	1980	6131
1957	6833	1965	8639	1973	9746	1981	5649
1958	6421	1966	8248	1974	7483	1982	6237
1959	7260	1967	8734	1975	6428	1983	6895
1960	7686	1968	8971	1976	7348	1984	6643
1961	7882	1969	8165	1977	6369		

Source: Timber tables from *Annual Abstracts of Statistics*, Central Statistical Office

Likewise, the 1957 figure is replaced by the mean of years 1955 to 1959, i.e.

$$\frac{7228 + 6829 + 6833 + 6421 + 7260}{5} = 6914.2$$

and so on.

Naturally we lose some points at the beginning and end (four in this case) but this may not be of great importance in a long series.

The line diagram (see Fig. 12.1) of the averages is seen to be much smoother than that of the original data (recall Section 3.6). However, a number of comments need to be made:

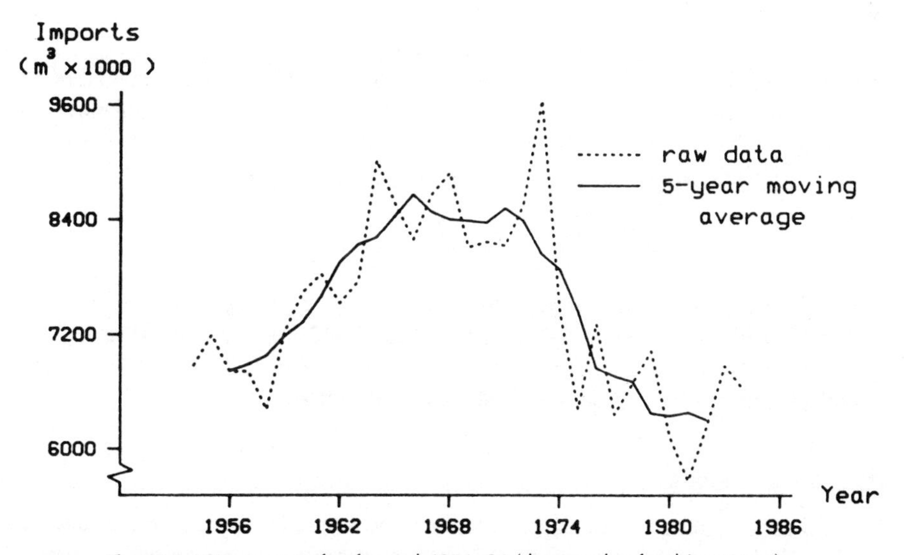

Fig. 12.1 UK imports of softwood 1954–84 (thousands of cubic metres)

1. Clearly the number of observations (the **period**) used to produce each average (five in our case) is a matter of judgement. If the period is too short, little smoothing will be achieved; if too long, the irregularities will be flattened out to an almost straight horizontal line and there will be considerable gaps at each end of the series.
2. If there are regular fluctuations (e.g. weekly, seasonal), these will be eradicated by using a period equal to the length of the fluctuation. On the other hand, if, for example, the data follow a 2-year cycle, a 3-year moving average will follow an inverse cycle since alternate averages will contain two lows and one high or two highs and one low.
3. Odd-length periods have the merit of being centred on the middle observation of the period; even-length periods provide a smoothed value that must be placed midway in time between the middle two observations of the period. If we are interested only in smoothing for pictorial representation, this does not matter. However, we often require smoothed values corresponding to the time points of the original observations yet have no choice but to use 12-month or 4-quarter averages to eliminate seasonal variation. The next subsection resolves this.

12.2.1 Centred moving averages

Suppose we have quarterly figures $y_1, y_2, y_3, y_4, y_5, \ldots$ Then the first 4-quarter moving average will be

$$\tfrac{1}{4}(y_1 + y_2 + y_3 + y_4)$$

corresponding to a time point midway between the second and third quarters, and the second will be

$$\tfrac{1}{4}(y_2 + y_3 + y_4 + y_5)$$

corresponding to a time midway between the third and fourth quarters. Hence an appropriate smoothed value to put in place of y_3 will be the average of these two, i.e.

$$\tfrac{1}{2}[\tfrac{1}{4}(y_1 + y_2 + y_3 + y_4) + \tfrac{1}{4}(y_2 + y_3 + y_4 + y_5)]$$

$$= \frac{y_1 + 2y_2 + 2y_3 + 2y_4 + y_5}{8}$$

which is just a weighted average (Section 3.1.5) using weights 1, 2, 2, 2, 1. The statistic would be called a five-quarter **centred moving average**. Although five different quarters are involved, each *kind* of quarter (1st, 2nd, 3rd, 4th) carries equal weight and so the seasonal effect is eliminated.

12.2.2 Forecasting

Clearly all manner of weighting schemes can be devised for special purposes. One system that gives rise to **exponentially weighted** moving averages is constructed so that greatest weight is given to the most recent observation, the weights applied to previous observations diminishing period by period by a constant factor, a. This is often used to predict the next observation using the most recent $(k+1)$ observations of a moderately stable series i.e.

$$\hat{y}_{t+1} = w_0 y_t + w_1 y_{t-1} + \ldots + w_k y_{t-k}$$

where

$$w_i = \frac{a^i(1-a)}{1-a^{(k+1)}}, \quad i = 0, 1, \ldots k$$

Commonly, a is chosen to be about 0.8 in which case, for a 5-period exponentially weighted moving average, we would have

$$\hat{y}_{t+1} = 0.2975y_t + 0.2380y_{t-1} + 0.1904y_{t-2} + 0.1523y_{t-3} + 0.1218y_{t-4}$$

Such a weighting scheme would not, of course, eliminate quarterly seasonal effects.

12.3 COMPONENTS OF TIME SERIES

Sometimes there is no coherent pattern in a time series and not much can be done by way of analysis other than elementary smoothing. Often, however, there are regular fluctuations relating to **seasonal variation** (S). These may be superimposed on an upward or downward overall **trend** (T). Any **residual variation** (R) is due to longer-term cycles and/or minor unexplained disturbances. Thus an observed value (Y) can be modelled (see Section 4.3) as

$$Y = T + S + R$$

and it is often useful to isolate these components. We shall use data on quarterly Scottish marriage rates shown in column Y of Table 12.2 and Fig. 12.2(a) to exemplify the calculation.

12.3.1 Trend

To isolate the trend, we must use a moving average of such period as to eliminate the seasonal effects. For these data we need a 5-quarter centred moving average so that the smoothed values coincide with quarters.

Thus

$$\frac{5.3 + 2 \times 6.8 + 2 \times 9.7 + 2 \times 6.8 + 5.8}{8} = 7.212$$

and so on, giving the trend T and hence the detrended series

$$Y - T = S + R$$

Notice how we can perform this calculation using 4-quarter moving totals, e.g.

$$(28.6 + 29.1)/8 = 7.212$$

Table 12.2 Scottish quarterly marriage rate (‰) 1977–84

	Y	M. Tot.	T	$Y-T$	S	$Y-S$	R
1977	5.3				−2.112	7.412	
	6.8				0.496	6.304	
	9.7		7.212	2.488	2.231	7.469	0.257
	6.8	28.6	7.312	−0.512	−0.615	7.415	0.103
1978	5.8	29.1	7.338	−1.538	−2.112	7.912	0.574
	7.1	29.4	7.312	−0.212	0.496	6.604	−0.708
	9.6	29.3	7.238	2.363	2.231	7.369	0.131
	6.7	29.2	7.188	−0.488	−0.615	7.315	0.127
1979	5.3	28.7	7.225	−1.925	−2.112	7.412	0.187
	7.2	28.8	7.275	−0.075	0.496	6.704	−0.571
	9.8	29.0	7.288	2.512	2.231	7.569	0.281
	6.9	29.2	7.362	−0.462	−0.615	7.515	0.153
1980	5.2	29.1	7.475	−2.275	−2.112	7.312	−0.163
	7.9	29.8	7.475	0.425	0.496	7.404	−0.071
	10.0	30.0	7.425	2.575	2.231	7.769	0.344
	6.7	29.8	7.375	−0.675	−0.615	7.315	−0.060
1981	5.0	29.6	7.262	−2.262	−2.112	7.112	−0.150
	7.7	29.4	7.088	0.612	0.496	7.204	0.116
	9.3	28.7	6.938	2.362	2.231	7.069	0.131
	6.0	28.0	6.838	−0.838	−0.615	6.615	−0.223
1982	4.5	27.5	6.800	−2.300	−2.112	6.612	−0.188
	7.4	27.2	6.775	0.625	0.496	5.578	0.129
	9.3	27.2	6.738	2.562	2.231	7.069	0.331
	5.8	27.0	6.725	−0.925	−0.615	6.415	−0.310
1983	4.4	26.9	6.750	−2.350	−2.112	6.512	−0.238
	7.4	26.9	6.775	0.625	0.496	6.904	0.129
	9.5	27.1	6.750	2.750	2.231	7.269	0.519
	5.8	27.1	6.788	−0.988	−0.615	6.415	−0.373
1984	4.2	26.9	6.912	−2.712	−2.112	6.312	−0.600
	7.9	27.4	7.012	0.888	0.496	7.404	0.392
	10.0	27.9			2.231	7.769	
	6.1	28.2			−0.615	6.715	

Source: Table A1.3 of the *Annual Report of the Registrar General for Scotland*, 1984

The trend is presented graphically in Fig. 12.2(b) and shows only minor fluctuations about a constant rate.

12.3.2 Seasonality

We are now left with seven years' worth of data containing only seasonal and residual effects. The four quarterly effects (which we are assuming to be invariant

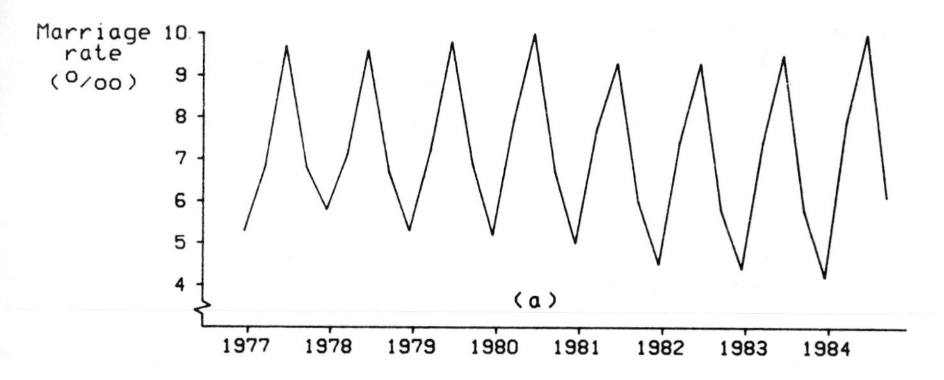

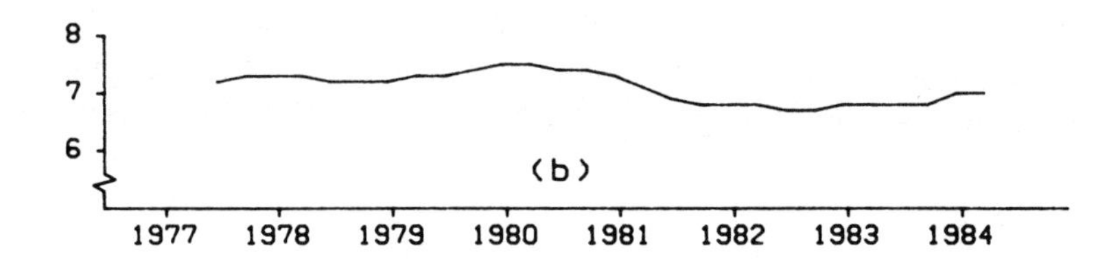

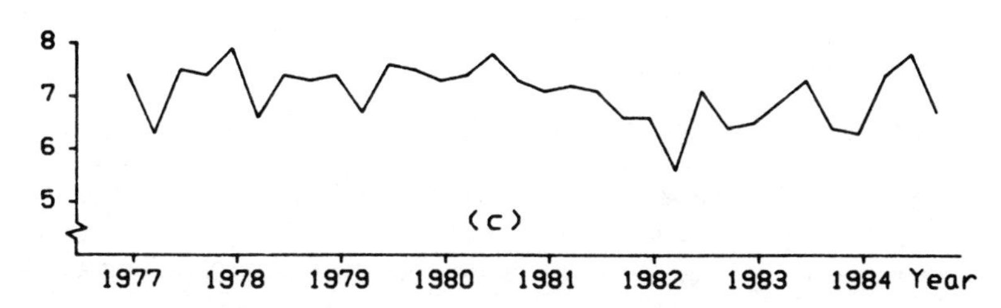

Fig. 12.2 Scottish marriage rates (‰) 1977–84: (a) row figures (Y); (b) trend (T); (c) de-seasonalized figures ($Y - S$)

during the time span of the data) can now be estimated from the averages of corresponding quarters over these seven years. Thus, for the first quarter, the effect is

$$\tfrac{1}{7}(-1.538 - 1.925 - 2.275 - 2.262 - 2.300 - 2.350 - 2.712) = -2.195$$

The full set is -2.195, 0.413, 2.148, -0.698. These add up to -0.332; to arrange that the average seasonal effect is zero, we should subtract $-0.332/4 = -0.083$ from each giving

$$S = -2.112, 0.496, 2.231, -0.615$$

We can now produce seasonally adjusted figures, i.e.

$$Y - S = 5.3 - (-2.112) = 7.412 \text{ etc.}$$

Figure 12.2(c) shows the de-seasonalized figures.

Except for the two quarters at each end, we can examine residuals

$$R = (Y - T) - S = 2.488 - 2.231 = 0.257, \text{ etc.}$$

and comment on any that are unduly large, indicating observations badly represented by the proposed model of trend and seasonality.

SUMMARY

Moving averages can be used to smooth time series. These can be centred if circumstances such as seasonal fluctuation require that the period be even. Time series may be decomposable into components representing temporal trend and seasonal variation thus permitting production of de-seasonalized figures.

EXERCISES

1. Table 1 of the *Digest of UK Energy Statistics* for 1987 gives the consumption figures for coal (millions of tonnes) and petroleum (millions of tonnes of coal equivalent: 1 tonne of petroleum = 1.7 tonnes of coal equivalent) for the years 1960–86 as in Table E12.1.

Table E12.1 Coal and Petroleum consumption in the UK, 1960–1986

Year	1960	1961	1962	1963	1964	1965	1966	1967	1968
Coal	198.6	193.0	194.0	196.9	189.6	187.5	176.8	165.8	167.3
Petroleum	68.1	73.4	80.8	87.9	96.1	106.2	114.8	122.6	129.4
Year	1969	1970	1971	1972	1973	1974	1975	1976	1977
Coal	164.1	156.9	139.3	122.4	133.0	117.9	120.0	122.0	122.7
Petroleum	139.6	150.0	151.2	162.2	164.2	152.2	136.5	134.2	136.6
Year	1978	1979	1980	1981	1982	1983	1984	1985	1986
Coal	119.9	129.6	120.8	118.2	110.7	111.5	79.0	105.3	113.5
Petroleum	139.3	139.0	121.4	110.9	111.1	106.1	135.2	115.0	112.6

(a) Superimpose the consumption figures in a line diagram.

(b) Add to your diagram the trend patterns derived from three-year moving averages. Do you feel justified in smoothing the figures across the sudden oil price rise of 1974 or the miners' strike of 1984?

(c) Could either of the trends be described as 'linear'? Summarize the consumption figures for coal and petroleum in a few sentences.

2. Table 1.1 of the *Monthly Digest of Statistics*, September 1987 gives the figures shown in Table E12.2 (£m) for UK Gross Domestic Product:

Table E12.2 UK Gross Domestic Product, 1982–87

Year	1982	1983	1984	1985	1986	1987
Quarter						
first	66 159	72 873	77 387	83 708	90 526	96 001
second	66 780	71 610	76 615	85 225	90 294	96 720
third	69 838	76 724	80 636	89 223	94 054	—
fourth	73 632	79 766	85 482	93 713	100 021	—

Determine trend and seasonal components for these data. Produce de-seasonalized figures and comment on the overall trend and any anomalous quarterly values.

3. Table 6.14 of the September 1987 *Monthly Digest of Statistics* gives the figures shown in Table E12.3 (in thousands of hectolitres of alochol) for home produced whisky:

Table E12.3 Home production of whisky, 1983–87

Year	1983	1984	1985	1986	1987
January	—	10.7	17.1	12.3	11.7
February	—	54.7	33.4	35.7	32.5
March	—	78.3	90.7	91.1	52.0
April	—	7.5	7.6	7.6	13.2
May	—	12.9	16.8	14.4	20.7
June	—	17.6	21.7	22.3	—
July	—	23.5	25.7	24.3	—
August	—	26.9	29.6	28.1	—
September	—	26.9	32.3	34.5	—
October	50.2	56.2	56.3	53.8	—
November	82.8	73.8	78.2	77.9	—
December	49.3	44.8	51.6	54.2	—

Source: HM Customs and Excise

Determine trend and seasonal components for these data. Comment on the seasonal fluctuation and any anomalous monthly figures.

13
Data analysis and the computer

13.1 THE COMPUTER

Nowadays almost all statistical analysis is accomplished with the aid of electronic computers. This situation is doubtless taken for granted by today's generation, yet, less than twenty-five years ago, leading statisticians were expressing concern lest interposition of the machine would so distance them from the data under consideration that the quality of analysis would suffer. What has certainly decreased over the years is the proportion of analyses made by qualified statisticians; the computer has brought to researchers an era of 'do-it-yourself'.

Unfortunately many still stand in such awe of the computer as to believe that any analysis achieved thereby must be correct. Nothing could be further from the truth. The computer can produce great volumes of rubbish at incredible speed. The quality of its analysis is entirely dependent on how intelligently it is used.

We digress a little at this point to introduce some of the jargon of the computer world.

(a) Hardware

The term **hardware** refers to the electronic and electromechanical bits and pieces that go to make up the machine. A computer is controlled by its **central processing unit** (CPU) operating on the **core** (or memory) which contains the instructions of the stored program (note the spelling) and also immediately required data. (The ability to store and hence repeat sequences of instructions is what characterizes the computer as distinct from the desk or pocket calculator.) The central processor also controls the transfer of information into and out of core as required by

1. input of data
2. output of results
3. intermediate storage of data on some form of **backing store** – magnetic tape magnetic disc or magnetic drum.

(b) Software

This term covers all the programs needed to accomplish an analysis. These include:

1. The **operating system** which arranges the sharing of the computer among several concurrent users, catalogues any data that are to be preserved in **files** for later use and provides various 'housekeeping' facilities helpful to the user;
2. **High-level languages** such as FORTRAN (FORmula TRANslation) or BASIC that are used by trained programmers;
3. **Specialized applications packages** that can be used by the relatively unskilled, e.g. SPSS (the Statistical Package for the Social Sciences).

In general, the full analysis of a set of data consists of a sequence of tasks from data validation, through the exploratory production of graphs and summary statistics to more complex techniques of estimation and inference.

Some software is therefore designed to run in **conversational mode.** The user, seated in front of a **video display unit** (VDU), interacts directly with the computer, inspecting the results produced by one request before deciding on the next. This approach is best suited to the study of small to moderate amounts of data but its merits have probably been exaggerated since even skilled statisticians need minutes rather than seconds to respond sensibly to output from all but the simplest analysis steps.

The alternative to this, essential for large volumes of data, is **batch working**. In this case, a sequence of tasks (possibly the complete analysis) is presented to the machine which then detaches the job from the user's control and queues it for running at a convenient time–perhaps overnight – the full results being returned to the user thereafter. Of course, any errors in the instructions submitted will prevent the machine from producing all that was asked and the job has to be corrected and re-run. Novice users who are most likely to have 'bugs' in their jobs find such delays particularly frustrating.

(c) 'Liveware'

This is the term coined to encompass all the personnel who surround (and fuss over) the modern computer – electronic engineers, systems programmers, applications programmers, duty advisers, information officers and those who give courses to would-be users.

13.2 THE ANALYSIS

Any computing activity can generally be subdivided into three parts:

1. **input** of the required data
2. performing a task – generally a **calculation** of some kind – using these data
3. presenting the results of the calculations to the user, i.e. **output**.

In many areas of the physical sciences, the second of these components is by far the most important. Two or three numbers might be required as input to a calculation lasting several hours, impossible to accomplish except by computer and producing only a small volume of output. In statistical analysis, on the other hand, the emphasis is on presenting to the machine information that is exceedingly diverse in nature and often very extensive and which can have a complicated, possibly hierarchical, structure. Great skill is also required of those who devise the layout of the results, since the output from the analysis many also be extensive, hard to assimilate and capable of misinterpretation.

We now go on to discuss the three aspects of analysis listed above.

13.2.1 Data preparation and input

As many have found, often only in retrospect, a good deal of thought must be given to the preparation of data for presentation to the machine.

Of course, some data are logged automatically and exist only in electronic form. Time series arising from monitoring a patient's heartbeat or from remote sensing some variable at the bottom of an oil well are obvious examples. Usually such data are so extensive that considerable compression or smoothing is a vital preliminary to analysis. In psychological experiments, data are often derived directly by presenting the experimental subject with the required test via the screen of a computer terminal; this is clearly most useful for visual perception studies.

Over the years, portable devices of greater and greater sophistication have been devised to facilitate easy and accurate recording of information at the point of collection. Use of a tape recorder in an interview is a crude example; automatic transfer of tree girths directly from the measuring tape to some magnetic storage medium is another. The main object of all such instruments is the reduction of transcription errors. Anyone who has tried to record information outdoors by pencil and paper on a wet and windy winter's day will appreciate this problem!

Optical scanning devices have become increasingly used for data capture either via some sort of bar coding system such as is commonly found at library issue desks, or by means of equipment able to read pencil markings at specified positions on a sheet of paper ('mark sensing'). The desirable goals of accurate reading of handwriting or indeed recognition of spoken words have almost been achieved.

For the time being, it remains true that most studies require some form of typing activity (key punching) in the preparation of data for the computer. Obviously, during this process, errors can be introduced due to misreading of badly written figures, misunderstanding about data layout or simple mistyping. Verification of the transcription phase is usually achieved by arranging that the data be punched twice by different individuals, the two versions being then compared by some automated means. Even this is not proof against all errors. If the volume of data will permit, 'calling over' the information as received by the computer against the *earliest* written documents can be helpful. In addition the computer itself can readily be made to perform certain checks.

The once ubiquitous 80-column punched card has now almost disappeared but, even though data are commonly input via a VDU, much of the flavour of the earlier era remains. This is particularly true in the case of questionnaire data from surveys (discussed more fully in the next chapter). Such data are usually extensive and a good deal of compression can be achieved by working in **fixed format** mode where each item of data for a sample unit occupies its own particular **field** (position) in the line being typed. The width of the field is determined by the maximal number of characters required for that variable. For greatest compression, each field must be as small as possible so that, for example, it may be acceptable to use a width of two for an 'age' variable by lumping together all ages of 99 and over. Use of fixed format input means that many coded questionnaire responses need occupy only a single character field. (Most often, a digit is used with 9 for 'don't known' or 0 for 'don't known' and 9 for 'other'.)

The alternative (and now much more common) **free format** mode requires separators such as a space or a comma to appear between items of data; this slows down data entry for a skilled keypuncher but is probably preferable for the non-expert since problems of field misalignment cannot occur. As an example, consider the layout of data on age of mother, weight (kg), average number of cigarettes smoked per day, number of days gestation, sex of baby, weight of baby (g),...:

Fixed format	*Free format*
3170304024712783...	31 70.3 40 247 1 2783...
2868700025423104...	28 68.7 0 254 2 3104...

A modern technique known as **forms mode entry** eliminates many data preparation problems and undoubtedly represents the way ahead. Basically, a general program, supplied with the data layout and validation requirements of the study, guides the keypuncher page by page through the questionnaire form (or similar document) in a helpful way so as to facilitate error-free transfer of the information and to give immediate warning of any data anomalies that can be detected as well as providing a mechanism for correcting mistakes. This process is usually mounted on a small computer dedicated to data entry, thus avoiding unnecessary use of the main machine; the transfer takes place only when the data are complete and free from error.

Even though more flexible ways of data preparation are welcome, past rigidity at least had the benefit of encouraging disciplined thinking about the relationship between the information to be collected and its analysis.

In all the discussion of this subsection, we are assuming that, after input to the computer, the data will form a named file in the backing store. From there it can be made available to the required analysis program(s) as often as necessary. Three decisions need to be made about how the information is to be held in the file.

1. **Character** versus **binary**. In the former, the data are held as coded characters exactly reflecting what would appear on a printed page. This is the usual method and amendments can readily be made to such a file if errors are detected. In a binary file, information has been converted to the machine's internal number representation and can be scanned much more rapidly by analysis programs. This is important for extensive data sets that are to be accessed often.
2. **Sequential** versus **random**. Data records held in sequential mode can be read only from first to last and all steps in any analysis run must be accomplished in a single 'pass' through the file. In a random (or **direct**) access file, records are individually identified by a **key** and can be read or re-read in any order. This is seldom needed for analysis purposes but random access mode is often used where the records are held over a long period and have to be updated with the passage of time.
3. **Flat-file** versus **database**. This is really an extreme form of the distinction made in 2 above. In a simple flat-file, the data are thought of as a two-way array of m variates (columns) measured on each of n units (rows), thus

$$\begin{array}{cc} & m \text{ variates} \\ & \begin{array}{cccccc} 1 & 2 & 3 & & & m \end{array} \\ \begin{array}{c} \\ n \\ \text{units} \\ \\ \end{array} \begin{array}{c} 1 \\ 2 \\ 3 \\ \cdot \\ \cdot \\ n \end{array} & \left[\begin{array}{cccccc} - & - & - & \cdots & - \\ - & - & - & \cdots & - \\ - & - & - & \cdots & - \\ \cdot & \cdot & \cdot & \cdots & \cdot \\ \cdot & \cdot & \cdot & \cdots & \cdot \\ - & - & - & \cdots & - \end{array} \right] \end{array}$$

This array is usually called the **data matrix**. From what was said in Chapter 4, the reader will appreciate that there could be several hundred variates and, of course, in the case of a census, n might be of the order of millions. Where data have a hierarchical structure or where units do not all have the same number of variates recorded (due, for example, to different numbers of visits to a clinic), the more elaborate structure of a database will be needed. Basically, this is achieved by the provision of software to facilitate the updating of and access to such data. One considerable advantage is that units satisfying specified conditions can be extracted for analysis without the need to scan the whole file; on the other hand, there are considerable overheads in setting up the database.

Since 1986, the Data Protection Act has required that all information about identifiable people stored by any electronic means is subject to a stringent registration scheme. Holders of such data must adhere to the very reasonable legal, moral and ethical principles laid down by the Act and the data must be held securely and confidentially. The Act is very wide ranging particularly in relation to the adjective 'identifiable' and subjects have considerable rights of inspection and amendment of their own records.

13.2.2 Statistical calculations

As we have implied above, statistical calculations are generally not as onerous as those of physics or chemistry. The greater benefit of the computer to statisticians is therefore that it does not become 'bored' by a large number of simple tasks. Data checking is a particular instance, e.g. in a study of obstetric history, the machine can be made to check that, for each woman,

total conceptions = total live births +
total spontaneous abortions +
total 'therapeutic' abortions

Computers are likewise good at some kinds of coding (e.g. social class from occupation), grouping (e.g. allocation to age class) and calculation of elapsed times (e.g. date of survey–data of birth). Any such calculation should be done 'manually' only if a useful check on data preparation would thereby be provided.

However, problems can and do arise even with statistical calculations, one of the commonest being the build-up of rounding errors. It must be remembered that computers are usually accurate to only six or seven decimal figures (whereas the average desk calculator can work to ten- or eleven-place accuracy). The following little example is instructive.

Example 13.1

We are to find the corrected sum of squares of three numbers using the formulae

(a) $$\sum(x-\bar{x})^2$$

(b) $$\sum(x)^2 - n\bar{x}^2$$

The computer we are using has a precision of six decimal figures.

Suppose the data are 1, 2 and 3.

For (a),

$$\bar{x} = (1+2+3)/3 = 6/3 = 2$$

$$(x-\bar{x}) = -1, 0, +1$$

$$\sum(x-\bar{x})^2 = (-1)^2 + 0^2 + 1^2 = 2$$

For (b),

$$\sum(x)^2 = 1^2 + 2^2 + 3^2 = 1 + 4 + 9 = 14$$

$$n\bar{x}^2 = 3 \times 2 \times 2 = 12$$

$$\therefore \sum(x)^2 - n\bar{x}^2 = 2$$

There is no problem with the calculation and both answers are the same because the maximal precision of any of the numbers involved is two, i.e. well within the capabilities of the machine.

Suppose, instead, that the data are

1001, 1002, 1003

For (a),

$$\bar{x} = (1001 + 1002 + 1003)/3 = 1002$$

and

$$(x - \bar{x}) = -1, 0, 1$$

so that, as before, the answer is correctly evaluated as 2 since no number exceeds six decimal figures of precision.

But consider what happens with (b). The squares of the data values require more than six decimal figures and must therefore be truncated by the computer. Thus

$$1001^2 = 1\,002\,001 \text{ which becomes } 1\,002\,000$$
$$1002^2 = 1\,004\,004 \text{ which becomes } 1\,004\,000$$
$$1003^2 = 1\,006\,009 \text{ which becomes } 1\,006\,000$$

Further,

$$n\bar{x} = 3 \times 1002 = 3006$$
$$\therefore n\bar{x} \times \bar{x} = 3006 \times 1002 = 3\,012\,012 \text{ which becomes } 3\,012\,010$$

and hence

$$\sum(x)^2 - n\bar{x}^2 = 1\,002\,000 + 1\,004\,000 + 1\,006\,000 - 3\,012\,010 = -10$$

We now have the embarrassing impossibility of a negative sum of squares and must conclude that method (b) is unsatisfactory for computer use even though it is the more practicable for hand calculation. The data used in the example may look unusual and have indeed been chosen for simplicity, but readers who suppose that these values are exceptionally troublesome are encouraged to experiment with others to confirm the reality of the problem.

However simple-minded the above example may appear, it underlines the need for caution in examining the results of statistical computations. There are undoubtedly items of software in use today that could produce negative sums of squares and many others giving answers that are undetectably wrong.

A problem peculiar to statistical analysis is the existence of *missing values* in data due, for example, to unexpected death of an animal or human subject, refusal of a respondent to answer all questions or even clerical carelessness. Sometimes the incompleteness need merely be noted and its extent reported, but other calculations are sensitive to incomplete data and the missing entries must be estimated. The way in which such imputed values are found varies from case to case and is seldom straightforward.

One statistical activity that often does give rise to lengthy calculations is simulation, a technique mentioned in Section 5.3 as a possible means of determining the form of complex probability distributions. However, simulation finds wider application nowadays in the many situations where it is necessary to model the behaviour of some process before its introduction or before a change is made. For example, a hospital, wishing to evaluate the possible benefits of introducing a pneumatic tube system for the transfer of blood samples between wards and laboratory, could use a computer to mimic the behaviour of the network under varying routine and emergency demand patterns and thus explore potential overload problems. Calculations of this kind can be particularly time-consuming (especially since there is generally no end to the scenarios that might be worth examining) and considerable ingenuity must be employed in writing software that will run efficiently.

Automated techniques for studying data have been provided in some software. These aids are generally of the 'try all possible combinations of...' variety and therefore demand excessive processing time besides inhibiting the use of common sense – the most valuable of all data analysis tools. However, current developments in the area of artificial intelligence do hold some promise of making available an efficient blend of human intuition and computer slog.

13.2.3 Output

In many ways, this is also a troublesome part of the analysis sequence. The difficulty faced by the software designer is to establish the correct level of detail to be provided, particularly in batch mode. It is tempting always to produce all the information that can be extracted by a particular statistical technique. This has the advantage of keeping the task request simple and the standardized layout of the output is easily scanned for points of interest by those familiar with the software. On the other hand, such an approach generally gives rise to very voluminous output neither easy to absorb from the screen of a VDU nor conducive to constructive contemplation when transferred to the continuous stationery produced by line printers. For users not expert in statistics, a great deal of the output will be meaningless or open to misunderstanding. No solution to this difficulty has been found and it remains a fundamental problem of 'do-it-yourself' data analysis. For, whereas a user of X-ray crystallography software is generally an X-ray crystallographer, most users of statistical packages are not statisticians.

Following the adage that 'a picture is worth a thousand words' one might suppose that graphical methods would provide an exceptionally useful form of output. For elementary exploration of a set of data, this is indeed true, but a great deal of work is needed to develop helpful graphical presentation of the results of the more sophisticated statistical analysis. The difficulty lies principally in that, whereas most scientists and engineers operate in a three-dimensional world capable of reasonable representation on a television screen, the multivariate nature of data analysis goes far beyond this into a space of m dimensions conceivable to the human mind only as an abstraction. In the area of graphical methods as well as in other areas of the interface between computing science and statistics, a great deal of exciting research work remains to be done.

SUMMARY

In the use of computers for statistical analysis, the greatest difficulty relates to data input and validation. Since the value of the resulting output is highly dependent on these components, they should be given early and careful consideration. The quality of even the latest software is not above reproach and much remains to be done to ensure that the computer is the servant rather than the master of its users.

EXERCISES

1. Experiment with the use of some statistical package to which you have access. (MINITAB is ideal for beginners.) Try repeating some of the exercises at the ends of Chapters 2, 3 and 4. Have you any criticisms of the output layout? How useful are the graphical facilities?
2. If you know how to write in some high-level language like Basic, Fortran or Pascal, construct programs to
   (a) find the mean and standard deviation of a set of numbers,
   (b) compute the correlation coefficient for n values of x and y,
   (c) (more difficult) de-trend and de-seasonalize a time series.

14
Social surveys

14.1 THE BASIC FRAMEWORK

In the widest sense, a survey consists of a number of organizational steps which need not occur in the order given below, which may often need to be retraced and which must usually be related to a preceding sequence of research or to comparable contemporaneous studies. Nevertheless, it is reasonable to pick out the following headings.

1. *Aims.* Are we interested only in summaries or descriptions or are hypotheses to be tested and inferences made? To what population are the results to apply?
2. *Review.* Available and relevant knowledge must be reviewed. Discussions with experts and interested bodies will be needed with possible consequent revision of 1.
3. *Research methods.* These must be designed or adapted from other studies *bearing in mind available resources*; a questionnaire of some kind will often be involved. A sampling scheme must be proposed (see Chapter 17). Provisional tables, etc., for the final report should be specified.
4. *Pilot study.* This is sometimes called a 'pre-test' and checks the feasibility of the scheme. A pilot is almost always essential. A revision of 3 may result.
5. *Staff training.* Personnel for interviewing, coding, etc., must be recruited and trained.
6. *Selection of sample.*
7. *Data collection.*
8. *Processing of results.* The information collected must be checked and corrected where possible. Coding will probably be required in preparation for computer analysis.
9. *Statistical analysis.* The interpretation of the output from the computer is basically as required by 1 but, almost always, new questions arise during the course of the study.
10. *Report.* This must be written at a level suited to the expected readership. Does the interpretation make sense? Is there support or contradiction from previous research? What might be done next? And so back to 1.

14.2 PILOT STUDY

Let us assume that we already know the size of sample, the type of respondent (child, housewife, air traveller, etc.), whether or not questions are to be asked on more than one occasion, whether there are likely to be seasonal fluctuations, whether the study is to be factual or attitudinal or both, and so on. The pilot study is a trial run on a small number of 'guinea pigs' who should be as like the true respondents as possible in age, intelligence, social class, etc. Naturally, most pre-testing effort should be put into areas thought likely to present problems and areas most crucial to the purposes of the survey.

After background reading and talks with experts (possibly the sponsors of the study) the initial pilot could be simply a number of lengthy unstructured interviews. If a formal questionnaire is to be involved one can test among other things:

1. the wording of any introductory letter
2. how to reduce non-response (e.g. colour of paper, age/sex of interviewer)
3. wording of questions (if ambiguities are found a further pilot is required to check the re-wording).

It may be possible as a result of the pre-test to change certain questions from free response to a more structured format. A re-pilot will be required.

For large, frequently repeated studies, each is the pilot for its successor but, on the other hand, changes militate against comparability over time.

14.3 METHODS OF DATA COLLECTION

The decision that must be made is between a postal questionnaire and some form of face-to-face interview.

14.3.1 Postal questionnaire

Arguments

For it is cheap (requires only printing of form, stamped addressed envelope, etc.)
hence a larger sample is possible for the same expenditure
the questionnaire form will almost certainly reach the desired respondent
a frank response to sensitive questions may be obtained

Against the questions must be simple because no further explanation is possible
the scheme is no use for respondents of low intelligence or limited educational background
'step-by-step' question responses cannot be obtained because the respondent usually reads all the questions before starting
the main problem is *low response rate*. This is typically 40% but may rise to 80% for groups especially interested in the study. The non-response is likely to be non-random so that bias is introduced; a

further complication is that the desired respondent may pass the form to someone he or she thinks is more interested.

Government surveys generally achieve a good response rate. It is usually found (from pilot studies) that 'personalization' of the letter of introduction is of no benefit, nor is use of a printed letterhead; neither the colour of paper nor the day (of the week) received seem to make any difference. Free gifts that can be claimed by return of the form can help. It has been found that stamped addressed envelopes give a better response than business reply envelopes (people do not steam off the stamps!). Reminders are sometimes helpful.

To maintain rapport in a postal survey we must always put ourselves in the position of the respondent. Apart from an explanation of the reason for the survey, how he or she came to be selected, a guarantee of anonymity, etc., each question must maintain the respondent's cooperation by treating him or her with courtesy and gratitude – a few 'pleases' and a 'thank you' are essential. It is worth avoiding official-sounding phrases such as 'marital status', 'length of service' and also slang unless it is really likely to be helpful. The questionnaire should be attractively printed and spaced to achieve a good layout and make things easy for the respondent.

Whatever the degree of response finally obtained it is good practice to investigate possible non-randomness and hence bias. Because it has been found that late returns come from people who are similar to non-responders, comparison of early and late returns may be useful. We can also relate the achieved sample to known facts about the population being sampled e.g. age/sex structure, geographical distribution.

14.3.2 Formal interview

(Face-to-face, possibly in the street but usually in the home, occasionally by telephone particularly for any follow up needed).

Arguments

For it is flexible in skilled hands, and allows explanations and probes pictures, lists, etc., can be shown (possibly in a randomized order) and samples can be handed over the inspection

some degree of immediate coding of responses may be possible

Against cost of training interviewers; supervisors are needed to give assistance to interviewers and check against possible interviewer dishonesty; travelling and subsistence must be paid even when the respondent proves to be unavailable or uncooperative interviewers may leave or become stale

collection of data may take a long time especially if the survey covers a wide geographical area there is always a danger of interviewer bias; interviewers may be tempted to express agreement with the respondent to maintain rapport; the interviewer's opinion may be conveyed by dress, accent, pauses or voice inflection in the reading of questions, thus revealing boredom, surprise etc. (Remember, interviewers are not a random sample of the general population.)

14.3.3 Self-administered questionnaire

The typical situation is a hospital out-patient department; the receptionist hands a questionnaire to the patient who completes it before or after treatment and returns the form there and then. Air travellers on a long flight provide another example of 'captive respondents'.
Arguments

For it is cheap because an already-paid official is used
sampling is usually accurate, e.g. every tenth patient
response rate is good because the respondent is captive and likely to be interested
personal contact and explanation are available where necessary but the risk of interviewer bias is small

Against the questionnaire has to be fairly short
response may be incomplete when the patient is removed by doctor or ambulance, or the plane lands, etc.

14.3.4 Group-administered questionnaire

Here, the respondents are typically a class of students or an invited audience. (An examination is a sort of group-administered questionnaire.) Questionnaire forms can be distributed and collected or the questions can simply be read out and the responses written on blank sheets of paper.
Arguments

For the respondents are again 'captive' and moderately interested
if the questions are read out, the time for each response is the same for all and can be predetermined
respondents can seek clarification

Against there may be contamination through copying, discussion or even any request for clarification.

14.3.5 'Hybrids'

All sorts of variations on these patterns and mixtures of the techniques are useful in specific situations. For example, the UK Census questionnaire is essentially hand-distributed, self-administered with an opportunity for clarification at collection time.

14.4 CONFIDENTIALITY

All information should be confidential in the sense that it cannot be traced back to an individual, i.e. anonymity must be guaranteed to achieve frankness. Thus, even for the Census, small random errors are introduced into district summary tables to prevent 'recognition' of unusual (e.g. 104-year-old) people. In general, therefore, it is best if a questionnaire does not request name unless a future follow-up is anticipated.

14.5 QUESTIONNAIRE DESIGN

Any but the simplest of questionnaires consists of a series of question sequences, each dealing with a different area of interest. (Occasionally it is useful to ask essentially the same question in different ways at different points as a check.)

14.5.1 Question sequence

Sequencing is at its most important in the interview situation where the respondent cannot 'look ahead'; but, whatever the method of data collection, it is sound policy for questionnaire designers to give some thought to this aspect both to clarify their own thinking and to simplify the respondent's task. The order of sequences should be considered first, then the question order within sequences. Within a sequence it is customary to move from the more factual to the more attitudinal questions, but whatever scheme appears natural and logical (to the respondent) should be used. In some cases, care must be taken not to put ideas into a respondent's head early in an interview lest these are reflected back at a later point.

For attitude surveys, a good order is:

1. questions to find out if the respondent has thought about the issue at all
2. open questions concerning general feelings
3. multiple-choice questions on specific matters
4. questions designed to find out reasons for views held
5. questions to find out how strongly views are held.

The **funnel effect** can be used to 'narrow down' a response step by step, e.g. 'Did you choose to study statistics to improve your job prospects?' is too leading a question. If this is indeed the respondent's belief, we want him or her to advance the idea spontaneously, so we might have: 'Had you any difficulty in making the choice of subjects that you study?'; 'Were short-term considerations such as time-tabling or published pass rates the only factors that influenced you?'; 'Do you feel your study of statistics will be of long-term benefit to you in the job market?'

A **filter question** is one that diverts respondents past a question sequence if it is irrelevant to them. Excessive use of this technique in a self-administered or postal questionnaire can irritate respondents so much that they give up.

14.5.2 Open questions

An open question invites a 'free' answer, recorded in full with the detail limited only by the space on the form. Coding is required but a selection of responses may be used verbatim in the final report to add interest. An open question gives great freedom to respondents but we may get just what happens to be uppermost in their minds. (This may be exactly what is wanted.) An open question is useful as a probe to clarify or expand the answer to a closed question, e.g. 'Why?'.

Open questions are easy to ask, difficult to answer and much more difficult to analyse. Coding is time consuming and usually cannot be planned before the start of data collection; when it can, **field coding** may be feasible. (With field coding, the interviewer asks an open question but codes the response immediately from a checklist not shown to the respondent.) Even for open questions, there will perhaps be loss of information if the interviewer does not use a tape recorder or due to the crudeness of coding required or the need to pool categories during analysis. We must naturally avoid any compression that causes bias. As usual, much of this can be sorted out in the pilot phase.

With questions on *behaviour* such as magazines read or television programmes watched the previous week, the open question produces lists that tend to be abbreviated due to forgetfulness. On the other hand, checklist questions sometimes inflate the response through confusion of 'last week' with 'previous week', etc. Diary techniques may be useful for longer-term studies provided they do not alter the respondent's habits.

14.5.3 Closed questions

A closed question is one with a fixed number of alternative replies. Because closed questions are easy and quick to answer, there can be more of them. The danger is that the respondent may become irritated by rigid categories constraining freedom of reply. There is also a risk that a closed question may bias response to a later open question by directing thought in a particular way.

There are various forms of closed question.

(a) Checklist

A list of options is presented which should include 'other' and 'don't know' categories. The 'other' category may require a 'please specify' probe. The selection of more than one category may be permitted.

(b) Rating

A rating gives a numerical value to some kind of judgement usually assuming equal intervals in the rating scale. Alternatively, a graphical scale can be used, e.g. 'Please mark your position on the following scale of satisfaction with this book'

Very dissatisfied |--------------| Very satisfied

For a rating scale, the number of classes varies from 3 or 5 to perhaps 10; it is usually difficult to make distinctions finer than a 10-point scale requires. An odd number of classes provides a neutral mid-point. The extreme categories tend to be under-used; we can try to make them sound less extreme or (particularly for a 10-point scale) we can pool responses from a group of end categories.

We have to make clear to the respondent what words like 'very', 'often', etc., are to mean. There are almost always problems of defining the end points of scales relating to, for instance, 'honesty'. Different respondents may use different **frames of reference** unless we tell them the purpose of the rating procedure.

Another problem is that of the **halo effect**; where a group of questions relates to different aspects of, for example, a particular statistics course, respondents with an overall feeling of like or dislike will tend to give high or low ratings to all the features. We can make it difficult for the respondent to do this by randomizing the direction of the successive ratings so that 'desirable' is sometimes on the left and sometimes on the right.

(c) Ranking

The respondent is simply asked to place a list in order. In survey conditions it is unreasonable to ask for a ranking of more than ten objects. If we must, we can use **classed ranks** – a hybrid between rating and ranking – in which we ask for assignment to a top class, second class, etc., there being relatively few classes. A large number of **preference pairs** gives the same kind of information as a rank ordering.

(d) Inventories

An inventory is a list about, for example, leisure interests or emotional feelings. Respondents are asked to tick those statements that apply to themselves or to give a mini-rating such as 'usually/sometimes/never'. Respondents should be told to read all statements before answering so that they have a 'feel' for such a rating scale. A variation is the **adjective checklist** in which respondents are asked to indicate any of a list of adjectives they feel apply to themselves or to others, or to a product, etc.

(e) Grids

A grid is essentially a two-way inventory allowing rapid collection of a great deal of information, e.g. 'Tick the textbooks you would use for each of the following topics'

	Graphical summaries	Index numbers	Hypothesis testing	Demography
'*The Valid Assumption*'				
'*Knowing About Numbers*'				
'*Fiddling With Figures*'				
'*The Meaning of Data*'				
'*Banishing Bias*'				

14.5.4 Dangers in question wording

Whether closed or open questions are used there are many problems of question wording.

(a) Vagueness

Consider, for example, the question 'Which textbooks have you read in the last six months?'. Does 'read' mean 'studied assiduously from cover to cover' or 'borrowed to look at one or two chapters' or 'glanced at in a library'? The purpose of the questionnaire will help us to tighten up the wording.

Social and/or geographical ambiguities can cause problems of interpretation even with factual questions, e.g. 'When do you usually have tea?' The word 'tea' could mean the beverage or a meal varying widely in size and time over different parts of the country and different social classes.

A particularly dangerous form of ambiguity is exemplified by 'Do you know if your mother ever had measles?' What will the answer 'No' mean in this case?

(b) Loaded words/phrases

These are emotionally coloured and may suggest approval or disapproval of a particular point of view, e.g. 'racist', 'aristocrat', 'deprived', 'ambitious'. One problem is that a word may be loaded in a special way for some part of the sample, e.g. 'disabled' is a much more acceptable term than 'crippled' for those in such categories.

In interviewing, different word stress can cause trouble through the interviewer's misunderstanding. Try asking the probe 'Can we eliminate football hooliganism?' with the accent on each of the five words in turn.

(c) Leading questions

These also suggest what the answer should be or indicate the questioner's point of view, e.g. 'Are you against excessive cuts in university funding?'. The term 'excessive', however it might be quantified, must invite the answer 'Yes'.

(d) Prestige bias

When asked in general terms, people claim to include a great deal of high-fibre food in their diets, to read only quality newspapers, never ever to drink and drive, etc. To avoid this we can emphasize the need for honesty, or provide a filter question so that a low-prestige answer seems equally acceptable. For example, instead of 'Have you watched any of the following television documentaries?', try the preliminary filter 'Do you ever watch television documentaries?'

Prestige bias can also result from questions introduced by, for example,

'Experts believe that ...'
'It is often said that...'

(e) Occupation bias

The question 'What is your job?' often produces considerable prestige bias (so that a 'plumber' becomes a 'heating and sanitary engineer') or vagueness, e.g. 'businessman'. This can be clarified only by requesting a brief job description.

(f) Ignorance/failure of memory

Respondents are always reluctant to admit such deficiencies; questions must be phrased so that they seem quite acceptable, e.g. 'Which brand of washing powder did you buy last?' might become 'Can you remember which brand...?' For lists, the 'don't know' category must always be made available.

(g) Embarassing questions

These ask about attitudes or behaviour that are not socially approved, e.g. drug abuse, truancy. The anonymity guarantee must be supported by an atmosphere of permissiveness in the question, e.g. someone may find it easier to admit that they have broken copyright law when faced with the options.

'1. never
2. once or twice
3. on many occasions'.

(h) Habit questions

Examples are:

1. 'When did you last...?'
2. 'How often in the last week...?'
3. 'How often do you... on average?'

1 and 2 may be subject to seasonal variation: 3 is vague about the period covered – the average may be changing.

(i) Over-long lists

In the interview situation, shyness or nervousness of respondents can cause them to forget items from a long multiple-choice list. Keep lists short or show the respondent a card with the relevant items.

There is always a danger of order bias – people remember and therefore choose items near the beginning or end of a list. Even in a list of two items, the second will be favoured. Either the list can be presented in random order to each respondent or each interviewer has the list in a different order.

(j) Double-barrelled questions

An example is 'Are you a vegetarian for ethical or health reasons?' Does 'yes' imply 'ethical' or 'health' or 'both'?

(k) Double negative questions

An example is 'Would you rather not live in a non-flouridated water area?'.

14.6 ANALYSIS

At *all stages* of a survey, but particularly analysis of the data, considerable cooperation is required among:

1. the sponsor
2. the interviewers and coders
3. the data preparation team
4. the statistician
5. the computer programmer.

Usually the sponsor or the statistician is 'in charge'. The analysis phase covers conversion of the information collected into a form suitable for computer input, validation, calculation of appropriate statistics, interpretation of these and preparation of a report.

The coding frame for each question should have been constructed at the pilot stage except for open questions where we may need to look at all responses to decide on groupings. The coding frames are collected together in a 'code book' that contains details of all codes, interviewer instructions, etc. Copies of the book will be required by all those connected with the survey and all copies must be kept up to date by the incorporation of any coding additions, new conventions and unresolved problems.

Coding can be done by interviewers to give them a break and to sustain their interest, but is more usually done by special staff. Either a whole form is completed by a single coder who then has an overall picture of the response and may thus spot inconsistencies, or each coder concentrates on all responses to a particular question, thus developing considerable skill at that particular task.

14.7 THE REPORT

Much of what was noted in Chapter 7 is relevant here. One has to assume that the report will be read critically (though this may be a rather optimistic view). Establishing the 'required' length at the outset will help to get the balance right. The title must be brief yet reflect the precise area covered by the report; a longer sub-title can be included if necessary. A useful framework is:

1. *Introduction*: outline of previous knowledge plus research team's initial perception of problem.
2. *Review* of the literature.
3. *Methods* used, described in sufficient detail to enable the research programme to be repeated, possibly elsewhere. Mention unforeseen problems.
4. *Results* section: should include all useful *new* information provided by the survey summarized by tables, graphs, etc.
5. *Discussion*: including any inferences resulting from statistical tests together with tentative explanations and unresolved questions.
6. *Summary*: sometimes placed at the beginning – often the only part read.
7. *Acknowledgements* of help in the form of advice or assistance in the conduct of the survey.
8. *Bibliography:* using either numbered references or the Harvard system, e.g.

Yates, F. (1981) *Sampling Methods for Censuses and Surveys*, 4th edition, Griffin, London.

Yates, F. and Anderson, A.J.B. (1966) A general computer program for the analysis of factorial experiments. *Biometrics*, **22**, 503–524.

With this system, references in the text are quoted as '... Yates and Anderson (1966)...' or '... Yates (1981, p. 123)...'

SUMMARY

Organizing a survey is a non-trivial exercise. Several methods of data collection are possible and a pilot study will be necessary to check that the operation runs smoothly. In particular, questionnaire design calls on many talents and the proposed format requires extensive testing in the pilot.

EXERCISES

1. Aberdeen Royal Infirmary patients are to be surveyed in a bid to provide a better service.
   The survey begins on Sunday and the first 10 patients to be discharged from each of 20 wards chosen at random will be given a comprehensive questionnaire to complete anonymously.
   The answers will then by analysed to find out what changes or improvements, if any, are thought appropriate.
   Patients will be asked such diverse questions as their first reaction to hospital, the food, ward temperature and how they were treated.

   Aberdeen Press & Journal, 22/1/88

   Discuss this study critically, considering
   (a) how the sample is to be chosen (Aberdeen Royal Infirmary is a general teaching hospital with about 650 beds in 30 wards.)
   (b) the advantages and disadvantages of the proposed study scheme
   (c) an outline of the required questionnaire.

2. A college department wishes to find the views of students who have just completed its first-year course on various aspects such as content, presentation, work load, etc., relative to other departments. Suggest an appropriate method of investigation, justifying your choice, and suggest any problems you foresee.

3. State any biases you would expect to find, and what measures you would suggest (either in the planning of the enquiry or in the statistical analysis) for remedying them in the following situations.
   (a) In a study of the use of British Rail for long journeys, a random sample of households is selected and each member is asked to list all railway journeys over 40 miles taken in the last 12 months.
   (b) In a study of the association between family size and university education, each member of a random sample of all who graduated from a certain university in the last 20 years is asked for information on the number of brothers and sisters he or she has.
   (c) In a study of consumer preferences for brands of tea, a random sample of telephone numbers is selected and a telephone enquiry made as to the brand in current use at each address.
   (d) In a study of the relationships between road accidents and the sizes and ages of cars involved, a hospital casualty department questions all accident patients about the cars in which they were travelling.

4. Explain fully how you would plan a study of the use of leisure recreational reading (i.e. not directly required by courses of study) among students of a college of higher

education. Mention the most important points about the questionnaire you would use, but do not show it in full.

5. Criticize briefly each of the following methods for evaluating viewers' appreciation of a television company's programmes:
   (a) tape recording interviews with passers-by at a busy street corner;
   (b) scrutinising letters in the press and letters received by the company from viewers;
   (c) door-to-door enquiries by trained interviewers in randomly chosen areas;
   (d) employment of a number of paid viewers who are required to report on reception, interference, etc., as well as on the programmes themselves.

15
Schemes of investigation

15.1 INTRODUCTION

Science often proceeds by observing the response to some experimental disturbance, reform or treatment. Denote this intervention by X. In the case of agriculture, biology, medicine and even psychology, formal laboratory or field experiments can usually be done; in the social sciences this is generally not possible but a good survey can often be thought of in quasi-experimental terms. We met this distinction between controlled and uncontrolled X in Section 6.1.

Because comparison is generally more important in science than the determination of an absolute value, we are generally involved in observing either the same sample at two different points of time between which X occurs, or two different samples at the same point in time, only one having been subjected to X.

15.2 PROBLEMS

What problems beset this process? (Because research on human beings presents the greatest difficulties we shall concentrate initially on this area.) To start with, we shall assume that the *same sample* of subjects is observed at times T_1 and T_2.

15.2.1 History (i.e. changes external to the subjects)

Any number of unknown external events may have affected the status, attitude, behaviour, etc., of a subject between T_1 and T_2. This is greater, the longer the interval between T_1 and T_2. In a laboratory context, the isolation of the respondent during a $T_1 - T_2$ period of a few hours should remove this effect; even changes such as sunlight may need to be eliminated. For longer intervals there may be a seasonal effect, e.g. depression in winter, anxiety as exams approach.

15.2.2 Maturation (i.e. changes within the subjects)

Between T_1 and T_2 a respondent will have grown older, possibly more tired, more hungry, etc. This includes the possibility that, for example, hospital patients or backward children could improve spontaneously even without X.

15.2.3 Mortality (a special case of the previous problem)

Some respondents may die or otherwise be lost to the study between T_1 and T_2.

Suppose that over a six-month period, we were investigating changes in attitude of a sample of *single* girls to contraception, some of the group might get married. Removal of these from the T_1 observations could bias the findings.

15.2.4 Changes in response at T_1 due to observation at T_1

The very fact that subjects know that they are being observed may change their responses. This is the so-called **reactivity** or Hawthorne effect, named after its discoverer.

15.2.5 Changes in response at T_2 due to observation at T_1

The T_2 observations may be affected simply by the fact that this is at least a second experience of being observed, i.e. there may be a **learning** effect due to T_1. It may be that questions asked at T_1 *cause* respondents to develop and shift attitudes during the T_1–T_2 period possibly through private discussion or, if the study is sufficiently large, public controversy. If possible, the T_1 observation should be of such a nature as to minimize reaction of this kind.

15.2.6 Changes in instrumentation

This relates to bias introduced by the use of different observers/interviewers at T_1 and T_2 or simply the same people becoming more skilled/bored, or the researcher may be tempted to use 'improved' scales of IQ, occupational prestige, etc., at T_2.

15.2.7 Selection bias

Suppose that on the basis of the observations at T_1, we select a group of children of low reading ability for experimental remedial education, X. Even if X is useless, a child's performance at T_2 cannot (or is unlikely to) deteriorate since it is already at an extreme low but, through chance variation, it may move upwards giving an overall appearance of improvement.

The data in Table 15.1 are simulated observations from a trivariate distribution in which the population means of all three variables are 55, the standard deviations are 15 and the correlations are $\rho_{yz} = \rho_{xy} = 0.5$ and $\rho_{xz} = 0.25$. It will be seen from the histograms, scatter diagrams (Fig. 15.1) and sample statistics that the random sample of 50 observations reflects these properties reasonably well.

Suppose we think of these data as representing the marks of 50 students in statistics examinations; the variable x, gives the marks at a Christmas examination and y gives the marks of the same students at Easter, the overall correlation being 0.523. If we consider the ten poorest students at Christmas (i.e. those with marks equal to or less than 40) we find an average increase in marks from 33.6 to 43.6. This observed 10-mark increase (with a standard error of 4.41) appears to indicate a real improvement.

Table 15.1 A random sample of 50 observations on variables x, y and z

x	y	z	x	y	z
28	29	40	54	56	62
29	37	61	54	42	61
30	65	50	54	48	48
31	44	53	54	37	56
31	33	46	56	48	63
32	45	55	57	69	59
35	32	46	58	64	59
40	36	37	60	54	67
40	73	77	60	52	53
40	42	51	60	63	85
42	57	46	62	54	58
44	40	35	62	69	56
44	59	47	62	81	64
46	59	70	64	40	59
46	68	59	64	65	59
48	35	42	65	67	76
48	57	62	67	43	45
49	58	61	70	79	59
50	30	46	72	78	48
50	36	29	73	58	58
50	39	61	73	57	62
51	42	38	74	61	59
52	57	60	87	76	43
52	45	35	88	55	48
54	46	82	92	70	70

Histograms

Class	x	y	z
20–29	XX	X	X
30–39	XXXXX	XXXXXXXXX	XXXX
40–49	XXXXXXXXXXX	XXXXXXXXXXXX	XXXXXXXXXXXX
50–59	XXXXXXXXXXXXXX	XXXXXXXXXXXXX	XXXXXXXXXXXXXXXX
60–69	XXXXXXXXXX	XXXXXXXXX	XXXXXXXXXXX
70–79	XXXXX	XXXXX	XXXX
80–89	XX	X	XX
90–99	X		

	$\bar{x} = 54.1$	$\bar{y} = 53.0$	$\bar{z} = 55.3$
	St. devn. = 15.1	St. devn. = 14.1	St. devn. = 12.0

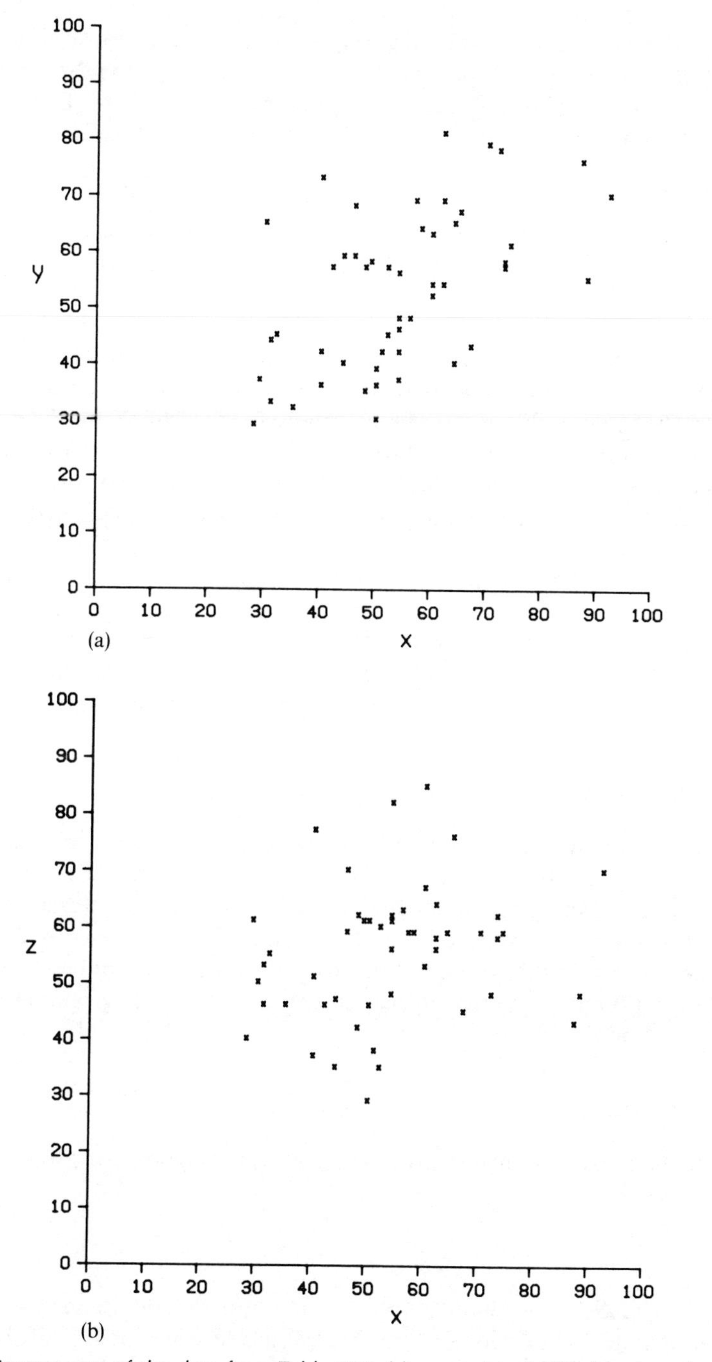

Fig. 15.1 Scattergram of the data from Table 15.1: (a) y vs. x ($r = 0.523$); (b) z vs. x ($r = 0.218$)

Thus, if the ten students had been given additional tuition, however ineffective, it would have *appeared* to be of benefit. All we have shown is that 'when you're at the bottom, the only way is up'. Clearly, therefore, there are dangers in the selection of extreme groups for special treatment since a change is *to be expected* even if the treatment is completely useless. This change is greater, the lower the correlation between the before and after measurements. For instance, if the Easter marks had been as in column z rather than y, showing a correlation with the Christmas performance of only 0.218, the average increase in marks of the lowest ten students would be 18 (standard error $= 3.63$) – an even more noteworthy 'improvement'.

Even when subjects are selected for experiment not on the basis of extreme scores in the attribute being studied but on the scores of some correlated attribute, the same problem can arise. Suppose for the same data the x-variable was a measure of reading ability in fifty 7-year old children, the y-variable a test of spelling ability and the z-variable the response to the same type of test given six months later. The average overall spelling mean is virtually unchanged. However, children with poor reading ability will tend to have lower spelling scores on the first occasion (since the correlation is 0.523) than on the second (for which the correlation is only 0.218). The result will be an apparent average increase in spelling ability for the ten poorest readers of 8 points (standard error $= 3.21$). If these ten poorest readers had been selected for remedial teaching during the six-month period a (probably) spurious improvement in spelling would be inferred.

15.2.8 Interaction between observation at T_1 and X

Those who have undergone measurement at T_1 may no longer be typical of the population in their reaction to X, the effect of which may be magnified or diminished. For example, someone who has just completed a questionnaire (at T_1) on nutrition may have a response to a film on food additives different from that which they would have had in the absence of the questionnaire. When the first observation is a *routine* measurement (e.g. a school examination) this problem is diminished.

15.2.9 Self-selection

Many of the above comments do not apply to a design of the form

$$\begin{array}{l} X \rightarrow S_1 \\ S_2 \end{array}$$

where S_1 and S_2 now relate to observations (made simultaneously) on two *different* samples of individuals, the first having received treatment X and the second forming the so-called **control** group. (The arrow simply indicates that X precedes the time of observation.) This is, of course, the special case of the design

$$X_1 \rightarrow S_1$$
$$X_2 \rightarrow S_2$$

where X_2 means 'no special treatment'.

However, a new problem arises – that of selection due to differential recruitment of those receiving or not receiving X. For example if, after a TV programme on violence in society, we sample to compare the opinions of those who watched it with those who did not on the question of bringing back corporal punishment, can we assume that the sub-populations of watchers and non-watchers are otherwise similar?

15.3 MATCHING

The main trouble with the scheme

$$\begin{aligned} X \rightarrow S_1 \\ S_2 \end{aligned}$$

is that we cannot be sure that an apparent X effect is not due to a differential influence of some extraneous factor(s) on the S_1 and S_2 groups. However, for such of these variables as we can conjecture may have such an influence, we may be able to **match** the experimental and control groups.

This matching (usually, but not necessarily, on a one-to-one basis) might be done **prospectively** at the start of the study, but, if that degree of control is possible, it is usually preferable to use randomization as discussed later (in Section 15.4). More usually, therefore, the matching takes place **retrospectively** by rejection of non-matched subjects at the time of selection of S_1 and S_2.

In such studies, the argument is cause-to-effect, i.e. we match a group that has received X with a control group and look to see if the observed responses in the S_1 and S_2 samples differ.

We could also set up an effect-to-cause study in which we match two groups for which the response measurements differ and look to see if they also differ in some previous X-type phenomenon. For example, in studying some congenital abnormality we could match a group of such children with a group of normal children and find out if, for instance, the mothers' smoking habits in pregnancy differed. In medical studies, this is sometimes the only ethical way to proceed; furthermore, the rarity of most abnormalities would demand a very large initial sample if the study were done by initial matching or by a cause-to-effect argument.

If we are interested in matching for age and educational level in the TV example of Section 15.2.9, two approaches are possible:

1. **Precision control:** a 1:1 (or $1{:}k$) pairing is made between S_1 and S_2 in respect of *all combinations* of the levels of age and educational attainment. The more extraneous variables involved, the harder it is to find matched subjects for many of the combinations;

2. **Frequency distribution control:** match each variable separately. This means we could have in the (unrealistic) case of two subjects per group,

   TV group: young and little-educated, old and well-educated
   non-TV group: young and well-educated, old and little-educated.

   We have matching for age and for education but it is possible that people who are older *and* well-educated have special opinions; such a person happens to receive X and may give rise to a spurious effect.

The object of matching is to increase the **internal validity** of the study, i.e. to try to ensure that any S_1–S_2 difference is due to X alone. This is done at the expense of the **external validity**, i.e. the applicability of the results to the population as a whole. The external validity may be less for at least two reasons:

1. The need to reject unmatchable subjects because they are in some rare category of the extraneous variables means that the sample groups do not truly represent the population;
2. By insisting that the two groups have the same distribution of some extraneous variable(s) we may be introducing unreal constraints, e.g. smokers and non-smokers do not have the same age distribution in the population.

15.4 RANDOMIZATION

The following techniques use randomization prior to the start of the study to remove most of the problems cited earlier. However, few *survey* situations allow such a randomization (R) between control and experimental (X) groups of subjects.

(a)
$$R\begin{cases} S_1T_1 \longrightarrow X \longrightarrow S_1T_2 \\ S_2T_1 \longrightarrow S_2T_2 \end{cases}$$

i.e. the sample of subjects is divided into two (not necessarily equal) groups by random allocation at the start of the study. Both groups are then observed (S_1T_1, S_2T_1), the experimental group receives X and the control group no treatment, then both are observed again (S_1T_2, S_2T_2). The correct analysis is to test the average change measured by $(S_1T_2 - S_1T_1)$ against the average change measured by $(S_2T_2 - S_2T_1)$.

If neither observers nor subjects know which subjects are receiving X and which are controls (a **double blind trial** – see Section 16.5), the interaction effect discussed in Section 15.2.8 will be removed.

If this is not possible we can achieve a similar result by the more complicated design

(b)
$$R\begin{cases} S_1T_1 \longrightarrow X \longrightarrow S_1T_2 \\ S_2T_1 \longrightarrow S_2T_2 \\ \qquad\qquad X \longrightarrow S_3T_2 \\ \qquad\qquad\qquad S_4T_2 \end{cases}$$

We can *measure* the effects of some of the problems listed earlier, by using only the observations at the second time point.

We can arrange these post-X observations as

	X	No X
With initial observation	S_1T_2	S_2T_2
No initial observation	S_3T_2	S_4T_2

The X effect is

$$\tfrac{1}{2}[(S_1T_2 + S_3T_2) - (S_2T_2 + S_4T_2)]$$

The effect of the initial observation is

$$\tfrac{1}{2}[(S_1T_2 + S_2T_2) - (S_3T_2 + S_4T_2)]$$

The interaction between initial observation and X is given by

$$\tfrac{1}{2}[(S_1T_2 - S_2T_2) - (S_3T_2 - S_4T_2)]$$

i.e. the difference between the X effects in the presence and absence of preliminary observation.

It will be seen that the selection difficulty presented by the design in Section 15.2.9 could be removed by randomization. Thus the following scheme is both useful and relatively simple:

(c)
$$R\begin{cases} X \longrightarrow S_1 \\ \qquad\quad S_2 \end{cases}$$

As against (b) we learn nothing about the (possibly interesting) history, etc., effects but the scheme is reasonably practicable.

SUMMARY

In any scientific investigation, the extent to which an observed difference is due only to some deliberate perturbation (internal validity) and the extent to which such a difference reflects a real difference in the parent population (external validity) is dependent on avoidance of a variety of pitfalls. These are most often found where human subjects are involved. Matching and randomization are two fundamentally important design techniques to enhance the validity and efficiency of a study.

EXERCISES

1. A social worker has developed a means of measuring 'emotional stability' in children by means of a short interview with the child. She now wishes to investigate whether the level of stability changes in the six months after children are moved from local authority

homes to foster parents. Write a report, warning her briefly of any problems and difficulties of interpretation that might occur and suggest how she might best proceed with the study.

2. New legal provisions on sex equality suggest that male nurses must be permitted to act as midwives. Since health service planners are concerned about the acceptability of this to mothers-to-be, they are to employ both male and female midwives for a trial period of 12 months in a certain area.

   How would you conduct a survey to investigate any change of opinion during the experimental period?

3. In recent years a policy of discharging long-stay psychiatric patients into the community has been introduced.

   Describe how you would conduct a study to investigate the well-being of such ex-patients.

16
Controlled experiments

16.1 TWIN STUDIES

One type of 'natural' matching has been used in some studies of identical (I) (about 1 in 5) and non-identical (N) twins. Twins, being of identical ages, are usually even better matched on environmental variables during upbringing than are siblings. Furthermore, I twins are genetically identical, coming from the same ovum, whereas N twins arise from two ova and are therefore no more similar genetically than ordinary siblings. We can sometimes use this to separate genetic and environmental effects – 'nature versus nurture'. The concordance between Is for some attribute (e.g. IQ) may be higher than that between Ns so that we can infer a genetic effect.

Suppose we measure smoking **concordance** as the percentage of cases where if one twin has the smoking habit then so does the other. If we find that the concordance is higher for Is than Ns, there could be a genetic component. But, for such an attribute, the difference could also be due to the greater psychological affinity between Is.

In the case of diseases, the effect of any genetic component is more clear cut. For instance, the Danish Twin Registry (using 4368 twin pairs of the same sex born from 1870 to 1910) found the following concordances.

	I	N
Rheumatoid arthritis	50	5
Tuberculosis	54	27
Rheumatic fever	33	10
Other acute infections	14	11

Because the sample is so large, even the last of these shows evidence of a genetic component.

However, notice that twins are unusual (about 1% of births) and this militates against external validity. Twins are more likely to be born to older mothers who already have several children; they are likely to be born prematurely and/or have difficult births. Some of these drawbacks to external validity diminish with the age of the twins being studied.

Of course, in a controlled experiment, where randomization is possible, we are perfectly entitled to combine the advantages of randomization and matching. This chapter is really about how to achieve this. The presentation is somewhat terse as the principles are more important than the details.

16.2 'BLOCKING'

As a first example, suppose we test the effect of some treatment, X, on pairs of identical twins, randomly allocating which is to receive X and which is to be the control. Genetic, sex, and environmental effects are eliminated in each within-twins difference and the mean of the these differences is a good measure of the X-effect even if the average response varies greatly over the twin pairs. We know this (Chapters 3 and 5) as a paired comparison study.

Such a design can be generalized to allow comparisons among more than two experimental treatments, though the rarity of human triplets, quadruplets, etc., means that the techniques are of use primarily in animal or plant studies. Suppose we have three drugs X_1, X_2, X_3 (or two drugs and a control) that are to be compared in their effect on rats. It will be easy to find litters of three (or more) and, even though such rats may not be 'identical triplets', the variability among siblings should be less than among rats in general. Thus:

1. comparisons of the drugs within litters reveal differences due only to the effects of X_1, X_2, X_3, i.e. there is good internal validity;
2. if the within-litter comparisons are replicated for a wide variety of rat parents, the experiment will possess good external validity;
3. if the treatments are randomly allocated (i.e. the firstborn rat does not always receive X_1) there should be no bias in the comparisons.

As another example, suppose that, in some sloping field, the effects of nitrogen (N), phosphate (P) and potash (K) on yield of barley are to be compared among themselves and with the yield from unfertilized (C) ground. The field is sufficiently large for six plots down the sloping length by two plots across; thus, there are 12 plots and we can replicate each of the four treatments three times.

A random allocation of the treatments might give rise to the following:

slope →

? dry	N	C	N	P	C	K	? wet
	C	N	P	K	K	P	

Notice that, by chance, the N plots tend to be near the top (? dry) end of the field and the K plots are towards the bottom (? wet) end so that we cannot be sure that any observed yield difference between N and K is not due to the moisture differential or to natural soil nutrients being washed by rain towards the low end of the field, etc.

Recognizing that plots of similar natural fertility, moisture, exposure, etc., are likely to be adjacent in the field, we can adopt a system of **local control** in which the area is conceptually divided into three **blocks** each of four contiguous plots:

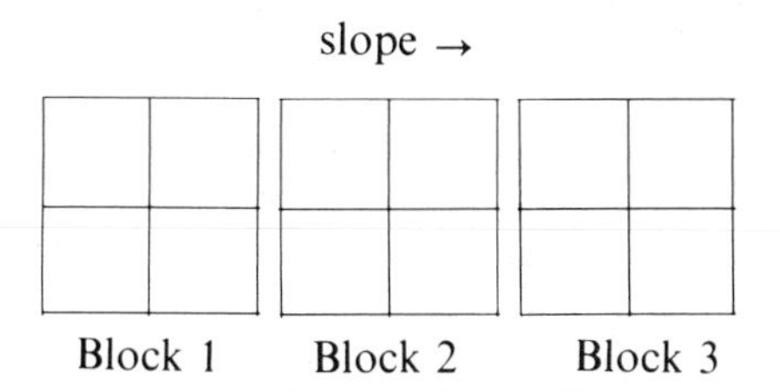

Within each block, the four treatments are allocated to plots at random so that we might have:

slope →

Block 1		Block 2		Block 3	
N	C	K	P	C	K
P	K	C	N	P	N

For this layout, the treatment comparisons *within* a block are less subject to unwanted external variation (both suspected and unknown). A measure of this variation will be available from discrepancies among the mean yields of the blocks.

In summary, to compare t treatments, we require a number of blocks each consisting of t experimental units; for one reason or another the units within each block are expected to behave reasonably similarly in the absence of any treatment, e.g. t mice from the same litter, t plots in the same area of a field. The t treatments are applied within each block by *random* allocation to the experimental units. Because each treatment occurs once and only once in each block, the response to a treatment averaged over the blocks is a valid measure of the effect of that treatment (and, less important, the block averages are valid measures of the block effects). The less the natural variation within blocks, the greater is the internal validity of the experiment; the greater the natural variability among blocks, the greater the external validity (though a single experiment is unlikely to be adequate in this respect); and randomization of the treatments within each block eliminates bias in the comparisons.

Such a scheme is called a **randomized block experiment**. More complicated designs are commonly used, particularly in agricultural research, but the basic principles are the same.

16.3 CONFOUNDING

One problem of the simple randomized block experiment arises when the number of treatments, t, is large. The straightforward scheme may be

1. *impossible*, e.g. we may be unable to get sufficiently large litters;
2. *unsatisfactory* since larger blocks will tend to have greater natural variability which runs contrary to the fundamental reasons for blocking.

One possible way round this is called **confounding**. Suppose we wish to study eight fertilizer treatments (X_1 to X_8) but can find uniform areas of ground (= blocks) for only four plots each. For just two blocks, we could have:

Block 1	Block 2
X_1	X_5
X_2	X_6
X_3	X_7
X_4	X_8

We can readily compare among X_1 to X_4 and among X_5 to X_8; but a comparison of any of X_1 to X_4 with any of X_5 to X_8 will be confounded with (i.e. inseparable from) the natural difference between the two blocks; this difference is unknown but could be large.

It is possible that these 16 treatment comparisons across blocks are of much less interest than the 12 within-block comparisons and we can be satisfied with what are effectively two separate four-plot experiments. Alternatively, if enough additional blocks can be found, it may be possible to allocate the treatments in such a way that any treatment comparison can be made within at least one block. Such a scheme is described as **partially confounded**. For example, we could compare three treatments in three blocks each of size two by:

Block 1	Block 2	Block 3
X_1	X_1	X_2
X_2	X_3	X_3

16.4 FACTORIAL TREATMENT COMBINATIONS

Let us return to our fertilizer experiment on barley involving N, P, K and C. You will notice how, on their own, these four **factors** (as we shall now call them) tell us nothing about how the fertilizers act *in combination*, e.g. does the effect of N differ according as P is present or absent?

To do this, we need to extend the list of treatments to give

C, N, P, K, NP, NK, PK, NPK

all eight appearing in each block (assuming there is no confounding). Such a scheme has a **factorial treatment structure** and is very commonly used in biological research.

Let the observed yield from the plot receiving a particular treatment combination be denoted by the corresponding small letter(s). Thus the overall effect of N versus C – the so called **main effect of N** – will be

$$\tfrac{1}{4}[(n-c)+(np-p)+(nk-k)+(npk-pk)]$$

averaged over blocks.

But we can also consider, for example, the difference between the effects of N in the presence and in the absence of P – the **NP interaction** (effect):

$$\tfrac{1}{4}[\{(npk-pk)+(np-p)\}-\{(nk-k)+(n-c)\}]$$

averaged over blocks.

This is, of course, the same as the differential effect of P in the presence and absence of N, i.e.

$$\tfrac{1}{4}[\{(npk-nk)+(np-n)\}-\{\{(pk-k)+(p-c)\}]$$

Example 16.1

Suppose that, for some block, we observe (in arbitrary units):

c	n	p	k	np	nk	pk	nkp
3	7	4	4	10	9	5	14

Then, for this block,

$$\text{Mean} = \tfrac{56}{8} = 7.0$$

$$\text{N main effect} = \tfrac{1}{4}[4+6+5+9] = 6.0$$

$$\text{NP interaction} = \tfrac{1}{4}[9+6-5-4] = 1.5 = \tfrac{1}{4}[5+3-1-1]$$

Factorial experiments are very efficient, since each experimental unit is used in the estimation of every main effect and interaction. The only problem is that the number of treatment combinations resulting is usually considerable so that large blocks must be tolerated or confounding introduced.

This last is less of a restriction than might at first be supposed. Remember that, although there are eight factorial combinations in Example 16.1, only seven comparisons are of interest, namely:

1. the main effects (N, P and K)
2. the two-factor interactions (NP, NK and PK)
3. the three-factor interaction (NPK)

Thus, if only four-plot blocks are available, we could have the following:

Block 1	Block 2
C	N
NP	P
NK	K
PK	NPK

The reader can check that the only one of the seven comparisons that is confounded is the NPK interaction and this is the least likely to be important. If it were thought to be worth estimating then a design involving more blocks and partial confounding would have to be used.

16.5 CLINICAL TRIALS

In the design of medical experiments (for the assessment of new drugs, for example) two additional problems generally appear. The first is the ethical difficulty associated with the need to give a patient the treatment that is believed to be best for him or her as an individual rather than what the statistician's random allocation might assign. We shall not pursue this further except to remark that a subject's consent to be part of an experiment is always obtained.

The second problem relates to biases introduced by the patient's or physician's preconceptions of the merits of a treatment. The danger is most marked where the measurement of improvement is subjective. For example, headaches and colds are very variable and virtually all disappear even without treatment. Furthermore, illnesses of a psychosomatic nature can be 'cured' by any treatment that the patient believes in. An objective physiological measurement is therefore preferable to asking people how they feel.

Where this cannot be done and the chance of subjective bias remains, experimental subjects are best divided into two groups – one group to receive the 'active' drug (D), the other to receive a **placebo** (P); this is literally a 'provider of comfort' and is an inactive imitation of the true drug that is identical in appearance, taste, etc. Until the experiment is over, neither the subject nor the doctor knows which is being administered in any particular case, the random allocation of patients to D or P being made at some remote centre.

Such a design is called a **double blind trial**. Clearly some treatments cannot be disguised (e.g. surgical operations) and we must also be sure that some side effect does not reveal the true drug to the doctor. (If the doctor knows whether a patient is receiving D or P, the trial is described as **single blind**.)

If patients entering an open-ended trial in order of referral to some clinic were to be assigned to D or P by, for example, tossing a coin, the numbers in the two groups could become unsatisfactorily discrepant. The reader can confirm from Table 5.1 that 10% of such trials would result in an imbalance of at least 20:10 after 30 patients. A technique known as **permuted block randomization** resolves this. The chronological patient sequence is divided into blocks of, say four. Within a block, six orders of equal D/P assignment are possible:

1	2	3	4	5	6
D	D	D	P	P	P
D	P	P	D	P	D
P	D	P	P	D	D
P	P	D	D	D	P

The decision as to which allocation is to apply to any block of four patients is simply made by random selection of a digit in the range 1 to 6.

SUMMARY

Matching can be used in twin studies or, more generally, in paired comparison experiments to reduce unwanted variation. The principle can be extended for larger blocks of inherently similar experimental units and can thus increase the sensitivity of an experiment to detect a real effect if it exists. Factorial treatment structure brings considerable advantages of interpretation but may also require over-large blocks so that confounding becomes necessary. Medical experiments must, if possible, be designed to eliminate psychological biases in patient and physician.

EXERCISES

1. A drug company is developing a new drug likely to be useful in the alleviation of migraines and other headaches. Unfortunately it may conceivably cause liver disease if taken over a long period, and this is to be tested by administering it to pigs. (Pigs are biologically very similar to humans.) Its merits as a headache cure are to be evaluated using the patients in two general practices.

   Advise on how these two aspects of the development programme should be conducted, indicating any reservations you may have about the methods or possible findings.

2. Distinguish the benefits of matching and randomization in designed experiments.

   Sixteen tomato plants are grown on a greenhouse bench in two rows of eight.

   Make a list of reasons why the yields of the plants will not be identical.

   Four growing regimes are under consideration involving two amounts of water supplied daily each with or without the addition of a trace element.

   How would you allocate the treatments to the plants if you were told that the temperature gradient along the bench was

(i) nil,
(ii) moderate,
(iii) considerable?

3. Using the method of comparing population proportions given in Exercise 5 of Chapter 5 and the fact that about one in five twins are identical, check the assertion that the Danish Twin Registry data given in Section 16.1 show evidence of a genetic effect even for 'other acute infections'.

4. Thirty elderly patients are available for a study of hip joint replacement. Twenty need replacement of both joints (bilateral); the remainder need only one. The question at issue is whether it is 'better' for the bilateral cases to have both joints replaced in a single operation or to have each joint treated separately in operations, say, six months apart.

   Two surgeons will perform all the operations. It is proposed to quantify the patients' recovery and rehabilitation by measuring the oxygen intake required during a walk of 100 metres ten days after the operation.

   Write a report commenting on the experiment.

17
Sampling schemes

17.1 INTRODUCTION

The simple random sampling scheme described in Chapter 1 is widely used and forms the basis of more sophisticated procedures that are particularly needed in the social sciences and medicine. In these circumstances we are usually investigating a finite population and therefore a **sampling frame** (i.e. a list of population members) may well exist. For some of this chapter we require additionally that such a list contains subsidiary information about each unit such as age and sex. Such attributes will certainly be known if the frame is derived from some kind of register, of patients, employees or students, for example.

A note about choosing the sample size is in order here although the decision about how large n should be is usually difficult. In general terms, some idea of the standard deviation of a variate has to be available from past experience and the acceptable width of confidence interval must be stated. This latter limits the permitted size of standard error and the relationship between standard deviation and standard error then allows determination of the appropriate n for that variable. For example, for a mean, we would have (see Section 3.6)

$$n = (\text{standard deviation}/\text{standard error})^2$$

The sample size must be the largest required for any of the required estimates or tests. In practice,

1. prior knowledge of basic variability is often inadequate particularly in a novel research area and/or
2. the value of n is limited by resources of time, money or experimental material.

The Greek letter ϕ is used to denote the **sampling fraction**, i.e. the proportion (n/N) of the population that is included in the sample.

17.2 SYSTEMATIC SAMPLING

Suppose the sampling fraction is to be 10%. Then, for **systematic sampling**, we select the sample members to be every tenth unit on the list starting from a randomly selected point in the first ten. In general, if the sampling fraction is ϕ, take every $(1/\phi)$th element. There are thus $(1/\phi)$ possible samples each defined by its first member and all equally likely if the starting member is chosen by selecting a random integer from the range $1, 2, \ldots, (1/\phi)$. (We are assuming for convenience that $(1/\phi)$ is an integer.)

If the sampling frame was constructed in such a way that the population members are listed in random order, then systematic sampling is equivalent to simple random sampling. The method is clearly convenient and cheap; for instance, it is very easy to sample every tenth new patient passing through a hospital clinic even though the population of patients is gradually changing.

The problem is that we cannot be sure that there is no hidden cyclic variation in the list sequence that could cause bias. For example, if we pick every tenth house number in a street then all those selected will be on the same side of the street; this could affect some aspect of the study. Likewise, if we sample every fifth fruit tree along a row in an orchard, we cannot be sure that the resulting gaps of, say, 20 metres are not coincident with some long-forgotten drainage scheme. However, the organizational advantages of being able to sample at a regular interval are so considerable that such anxieties are often set aside.

17.3 SIMPLE RANDOM SAMPLING FROM A SEQUENCE

Simple random sampling can be achieved even when no frame is available but where the population members present themselves one at a time as potential sample members. The penalty is that the procedure for determining whether or not a particular unit is to be in the sample is a little more complex than with systematic sampling or straightforward random selection of n out of N.

Let us suppose that the kth member of the population sequence is being considered for sample inclusion and that thus far in the sampling process, we have obtained n_k of the n sample members that we require. We consider three possible methods. Each requires us to be able to produce a (pseudo-)random value, μ_k equally likely to lie anywhere in the range 0 to 1. (Some pocket calculators will do this; otherwise one obtains a random integer of 5 or 6 digits from a random number table and precedes it with a decimal point.)

Method 1

Simply select unit k for the sample with probability ϕ, i.e. generate a random μ_k in $(0, 1)$ and if μ_k is less than ϕ, include this unit in the sample. The only disadvantage is that the final sample is unlikely to have exactly n members though this may not be very important.

Method 2

Select unit k for the sample with probability $(n - n_k)/(N - k + 1)$, stopping once the desired sample size has been attained, i.e. when $n_k = n$. A few moments' reflection, perhaps considering explicit values for N and n, will convince the reader that the final sample size must be n.

Method 3

This method can be used even if N, the size of the population, is not known in advance. It is most useful where records are to be sampled by computer from an unknown number held on, say, a magnetic tape. When the kth record is encountered, a random value, u_k, in the range 0 to 1 is generated; the final sample is to consist of those records associated with the n smallest values of u produced. As the sampling proceeds, therefore, a continual shuffling of proposed sample members into and out of an auxiliary subfile is required, a procedure ideally suited to the computer. The sample size is, of course, exactly n.

Although it may not be immediately obvious, all three methods produce simple random samples, i.e. every possible sample of size n out of a population of N has an equal chance of selection.

17.4 STRATIFIED SAMPLING

It will surely have occurred to the reader that a simple random sample, however 'correctly' taken, can be unrepresentative of the population from which it is drawn. Thus all members of a simple random sample of 30 pupils from a mixed school *could* be girls. It is an unlikely occurrence but it could happen and we would feel concerned if it did; indeed, we would probably be unhappy with any sample where the sex ratio differed greatly from that of the whole school. Likewise, we might be disturbed if the social class distribution in the sample turned out to be very unlike the overall pattern.

If the sampling frame contains information on important qualitative attributes such as sex, social class, etc., that split the population into subpopulations or **strata**, then we can enhance representativeness by constructing the overall sample from the amalgamation of simple random samples from within every stratum, e.g. within each sex stratum or within each sex-by-class combination. By this means we may greatly enhance the *external validity* of the study (see Section 15.3). Furthermore, in practice, we are often interested in making comparisons among the strata so that it is sensible that each be well represented.

Another reason for stratification is akin to that for local control in field experimentation, i.e. variance reduction. (Indeed stratification is to the survey what local control is to the experiment.) For, suppose we know that the main variable of interest in the study varies much less within strata than overall, then much more precise estimates are possible for this variable. The apparent advantage of variance reduction is diminished if, as is much more often the case with surveys than with experiments, many somewhat disparate characteristics are under review rather than just a few correlated variables.

17.4.1 Allocation of sample units to strata

Once a decision to adopt a stratified sampling scheme is made it becomes necessary to decide how the n units in the sample should be allocated to strata.

Suppose that the ith stratum contains N_i population members ($\sum N_i = N$) and we are to select n_i sample units ($\sum n_i = n$) from it. Two approaches are possible:

(a) Proportional allocation

In this case the overall sampling fraction, ϕ, is applied to each stratum, i.e.

$$\frac{n}{N} = \phi = \frac{n_i}{N_i}$$

Thus

$$n_i = N_i \times \frac{n}{N}.$$

The n_i values thus calculated will generally have to be rounded and adjusted so that their sum is n.

(b) Optimal allocation

This procedure assumes that, from past experience (possibly a pilot study), we know the standard deviation, d_i, of the main variable of interest within stratum i. Then the sampling fraction in stratum i is made proportional to d_i, i.e.

$$\frac{n_i}{n} = \frac{N_i d_i}{\sum(N_i d_i)}$$

(This assumes once again that there is a single variable of primary interest.) The more variable strata will therefore be more intensively sampled, leading to estimates of the population mean or total for this variable that are 'optimal' in the sense of having smallest possible variance.

We can perhaps clarify the last two methods by means of a simple example.

Example 17.1
Suppose there are three strata as follows:

	Stratum i			
	1	2	3	
N_i	100	200	300	$\therefore N = 600$
d_i	4	3	2	

We require a sample of 100, i.e. an overall sampling fraction of $\phi = 1/6$.

Then proportional allocation determines the strata sample sizes to be

$$n_1 = \frac{100}{6} = 17 \text{ (rounded)}$$

$$n_2 = \frac{200}{6} = 33 \text{ (rounded)}$$

$$n_3 = \frac{300}{6} = \underline{50}$$

$$100 = n$$

On the other hand, optimal allocation would give

$$\sum N_i d_i = 100 \times 4 + 200 \times 3 + 300 \times 2 = 1600$$

so that

$$\frac{n_1}{100} = \frac{400}{1600} \qquad \frac{n_2}{100} = \frac{600}{1600} \qquad \frac{n_3}{100} = \frac{600}{1600}$$

i.e.

$$n_1 = 25$$

$$\left.\begin{aligned} n_2 &= 37 \text{ (rounded)} \\ n_3 &= \underline{38} \text{ (rounded)} \end{aligned}\right\} \text{or} \left\{\begin{aligned} 38 \\ 37 \end{aligned}\right.$$

$$100 = n$$

17.5 MULTISTAGE SAMPLING

Multistage sampling is used in circumstances where a hierarchy of units exists. For example, we could have a number of herds of cattle; these herds would be called **primary sampling units** and the individual cattle are **secondary sampling units**. Any number of levels is possible but, in what follows, we shall restrict discussion to such two-level hierarchies. The important point is that the two-stage sampling process involves a sample of the primaries from each member of which is taken a sample of secondaries. It is the lower-level sampling units that are of interest.

It is usually the case that the primaries are geographically distinct and, since not all are to appear in the sample, there is a clear administrative advantage over a sampling scheme involving the whole population of secondaries. For example, we might have 10 000 cattle distributed over 100 farms; a simple random sample of 200 cattle is likely to involve most of these herds, resulting in a great deal of travelling by the investigators. There is an obvious reduction in travel if we sample 10 herds and then 20 animals within each. A further gain is that we do not need a sampling frame for the 10 000 cattle but only for each of the 10 herds; such a frame could be constructed on the investigator's arrival at a farm.

17.5.1 Cluster sampling

A special case of this two-stage approach requires that a complete census be taken in each of the primaries selected; this is called **cluster sampling**. In our example, we would consider all animals in each of the herds sampled. It is because the individual animal is the unit of study that this is a cluster sample; if herds were the unit of study we would simply have a simple random sample of herds.

This distinction is of some importance. A random sample of herds might indicate that 40% of them are grazing on marginal land; but if animals were the unit of study we might find only 20% of animals to be on marginal grazing, i.e. the 40% of herds are the smaller ones accounting for only 20% of the cattle.

The clusters need not be natural. They can be defined arbitrarily for the study, for example, by imposing a grid on a map, in which case the technique is called **area sampling**.

17.5.2 Two-stage sampling schemes

If we are to have a true two-stage sample, i.e. a sampling scheme for the secondaries within the sampled primaries, we have to face the fact that the primary sampling units can contain very different numbers of secondary sampling units. However, we require that every secondary sampling unit has an equal chance of inclusion in the sample. Two possible ways of doing this are now described. To exemplify these let us consider just three herds of cattle, A, B and C, and assume for simplicity that only one herd is to be sampled. The numbers of cattle in these herds are

	A	B	C
nos. of cattle	180	90	30

Method 1

We could simply choose the herd at random (i.e. with probability of 1/3 that any given herd is selected) and then use the same secondary sampling fraction for the animals whichever herd is involved, say $\phi = 20\%$. The probability that any particular animal is included in the study is thus $1/3 \times 1/5 = 1/15$, i.e. the probability of its herd being the one selected times the probability of that animal being then sampled at the secondary stage. Although this method fulfils the requirement that all animals are equally likely to be part of the study, we have no control over sample size; if herd A is chosen then there will be 36 cattle in the study whereas for herd C there will be only 6.

Method 2

We could sample the herds with probabilities proportional to their sizes, i.e.

$$\text{Prob. A is sampled} = \frac{180}{180 + 90 + 30} = \frac{180}{300} \; (= 0.6)$$

$$\text{Prob. B is sampled} = \frac{90}{300}\ (=0.3)$$

$$\text{Prob. C is sampled} = \frac{30}{300}\ (=0.1)$$

Then, whichever herd is sampled, a fixed number of cattle, say 20, will be chosen at random; the number of secondary sampling units is thus predetermined. Furthermore the probability that any animal in A is selected must be $180/300 \times 20/180 = 1/15$; for an animal in B the probability is $90/300 \times 20/90 = 1/15$; and likewise, for C, the probability is $30/300 \times 20/30 = 1/15$. The equiprobability requirement is therefore satisfied.

Although this second method is the more complicated to administer and requires that we have at least a rough idea of the size of the primary sampling units, it has some advantages that become obvious when we recall that several primaries are generally sampled:

1. It seems reasonable that larger and therefore, in some sense, more important primary sampling units should have a greater chance of inclusion in the study than small ones;
2. If interviewers or investigators are each allocated the same number of primary sampling units, they will all have the same workload;
3. If it is later decided to compare primaries, then equal-sized samples from each are likely to give greatest precision.

But how are we to sample the primaries with probability proportional to size? Suppose we now consider six herds from which two are to be sampled in this way. The numbers of cattle and the cumulated numbers are as follows:

Herd	Nos. of cattle	Cumulative nos. of cattle
A	180	180
B	90	270
C	30	300
D	75	375
E	67	442
F	144	586

All that is necessary is to select two numbers from random number tables in the range 1 to 586. For example, suppose we choose 289 and 423; then 289, being in the range 271 to 300, corresponds in the cumulative total to herd C; likewise 423 corresponds to herd E.

The difficulty is that we are quite likely to pick the same herd twice and if that were to happen we would have to sample it twice. An alternative selection method which generally avoids this problem is to take a systematic selection of 2 from the range 1 to 586. Thus we pick a random number in the range 1 to $586/2 = 293$. Suppose we get 204, giving herd B; the second of the systematic sample will be $204 + 293 = 497$ which gives herd F. By this means we always obtain two different primaries unless one of the primaries accounts for more than half the secondaries, in which case that primary has to appear at least once.

17.6 QUOTA SAMPLING

This technique is particularly associated with street interviewing for public opinion and market research surveys as an alternative to calling on randomly selected households. A stratification requirement is specified, such as age-by-sex, and the numbers to be sampled from the strata as discussed in Section 17.4.1. Interviewers are then assigned **quotas** for interview, e.g. '3 men aged 20–29, 4 women aged 45–64,...' and they choose these sample units in any convenient way, usually by judging that an approaching individual fits one of their required categories and confirming this before the interview. The quotas, combined over interviewers, thus form a *non-random* stratified sample.

It is worth noting that a matching problem arises akin to that discussed in Section 15.3; the sample can be selected to mimic the population distribution of stratifying variables either individually or in combination. The first is clearly easier but less likely to be representative.

Quota sampling is obviously less costly than a simple random sample partly because no sampling frame is needed and no call-backs to unlocated sample members are involved. The greatest advantage, however, is speed so that an up-to-date measure of opinion on, say, a recent government action is rapidly available.

Against this must be set a non-response about twice that for a random sample involving house calls. Bias may be introduced by the time and place of the quota selection unbalancing the sample in respect of some unconsidered attribute such as employment status. The need to make subjective judgement of, say, age before approaching a potential respondent tends to produce ages unevenly clustered around the middle of the age ranges which must, in any case, be broad to make the technique feasible. Above all, there is no easy way of checking on the honesty of an interviewer who might, at worst, sit comfortably at home inventing questionnaire responses.

17.7 STANDARD ERRORS

Calculation of standard errors of sample statistics is generally more complicated for the sampling schemes described in this chapter than for the simple random sample. Even for this latter, the standard formulae must be amended if ϕ is anything over, say, 0.1. The non-randomness of the quota sample (and, to some extent, the systematic sample) prevent computation of any measure of variability.

SUMMARY

Although systematic and quota sampling provide samples that are not truly random, they can be used in circumstances where no sampling frame exists. If, on the other hand, a frame contains additional information about the population members, stratified sampling can improve both external validity and precision. Where a hierarchy of units exists, multistage sampling is appropriate and often has administrative advantages.

EXERCISES

1. The tutorial groups for students taking one of three courses are arbitrarily labelled A, B,..., P and the numbers of students in the groups found to be as follows:

Course										
1	Tutorial	A	B	C	D	E	F	G	H	Total
	Nos.	11	16	13	16	19	6	8	15	104
2	Tutorial	I	J	K	L	M				
	Nos.	11	11	13	10	10				55
3	Tutorial	N	O	P						
	Nos.	11	14	12						37

Explain *in detail* how to take each of the following samples of these students:
(a) a simple random sample of 45 students,
(b) a systematic sample of 28 students,
(c) a sample of 25 students stratified by course; sample sizes in strata are to be proportionally allocated and a simple random sampling scheme used,
(d) a cluster sample using tutorials as clusters; four tutorials are to be selected with probability proportional to cluster size, two from course 1 and one from each of the other courses.

2. You are asked to plan a survey to investigate the consumption of sweets by school children. Write a preliminary memorandum, embodying your proposals for a sampling scheme with a consideration of possible frames and indicating any difficulties of execution or interpretation that occur to you.

3. To investigate the attitude of residents in a residential area of a city to a number of civic amenities, an agency decides to sample 200 adults from the area. It proposes to do this by taking a random sample of 200 households, obtained from a listing of all households in the district and then selecting at random one adult from each such household. The sample members will then be interviewed in their homes during the evening.
Comment on this study proposal, discussing any possible sources of bias likely to be

encountered, as well as such problems as non-response. What improvements to the scheme can you suggest?

4. The households in a town are to be sampled in order to estimate the average value per household of assets that are readily convertible into cash. Households are to be divided into two strata, according to rateable value.

   It is thought that for a house in the high-rate stratum the assets are about four times as great as for one in the low-rate stratum, and that the stratum standard deviation is proportional to the square root of the stratum mean. It is also known that the low-rate stratum contains two-thirds of the population of households.

   Resources are available for a sample of 300 households. Distribute this number between the strata, using

   (a) proportional allocation,
   (b) optimal allocation.

   Indicate which method you consider to be more suitable, justifying your choice.

18

Longitudinal studies and interrupted time series

18.1 PREVALENCE

When the members of a sample are observed once at a single point of time, the study is termed **cross-sectional**. Suppose we are interested in deafness defined as the inability to hear some standard sound through headphones. From such a snapshot study, we can estimate the **prevalence** of deafness in, for instance, different age/sex categories. The prevalence of a state is defined as the proportion of a population so categorized at a given point of time regardless of when those affected entered the state.

18.2 INCIDENCE

Of greater interest is **incidence** – the probability of entering the state in a certain interval of (usually) age. To determine, say, the incidence of deafness in men between 50 and 55 years of age we really need more of a cine film approach. If we were to attempt an estimate from a cross-sectional study, it would have to be

$$\left.\begin{array}{l}\text{Prevalence of deafness in 50–55 age range}\\ -\text{ Prevalence of deafness in 45–50 age range}\end{array}\right\} = x - y$$

But this ignores the possibility of short-lived deafness, i.e. not all of y will be included in x so that the estimate will be too low. Such a bias may be reinforced by the fact that those who are deaf in the 45–50 age range are less likely than their fellows to survive to the following age group (because the deafness is related to some general malady or because hearing deficiency increases their risk of fatal accident). A further complication is the possibility of **secular drift**, i.e. prevalence changing with the passage of historical time. This, however, is unlikely to be a serious problem with five-year age bands. A cross-sectional study might supply some true incidence information if subjects were asked when they first noticed the onset of the condition of interest. However, unless an accident or acute illness was associated with onset, memory problems are likely to confuse the result. For example the gradual deterioration often associated with deafness will make it impossible to determine the point at whcih the required criterion of deafness was first satisfied.

A **longitudinal study** involves a pattern of the form

$$\cdots O_1 \; O_2 \; O_3 \; O_4 \; O_5 \cdots$$

where the *same* sample (sometimes called the **panel**) is observed (O_i) at (usually) equal intervals of time – say five years. Some of the problems noted in Chapter 15 can therefore appear. Sometimes 'continuous' monitoring is possible in the sense that a change of state triggers an observation; this can be the case in, for example, psychiatric follow-up studies.

Returning to our simple example of male deafness, for n men aged 45–50 on a previous survey five years earlier and now, therefore, aged 50–55, we might have numbers such as:

		Current survey (50–55)		
		deaf	not deaf	
Previous survey (45–50)	deaf	a	b	
	not deaf	c	d	$(a+b+c+d=n)$

The number, b, in the deaf→not deaf cell tells us how many revert to normal hearing; the estimated probability of reversion is thus $b/(a+b)$. Prevalence of deafness in the 50–55 age range is given by $(a+c)/n$. This could be compared with an estimate of the same prevalence from some previous survey to give a measure of secular change. The incidence is, of course, simply given by c/n. This simplified presentation takes no account of subjects dying or otherwise being lost to the study between the two observation points.

18.3 CROSS-SECTIONAL VERSUS LONGITUDINAL SURVEYS

Cross-sectional studies take place at a single point of time (although the 'point' may in practice cover several months). We have to assume that this point is in some sense typical. If comparisons of prevalence are to be made with other populations or other times, there must obviously be reasonable similarity in the research methods used. However, for a cross-sectional study, a body of field workers, equipment, working accommodation, etc., must be available for a relatively short time on a one-off basis. Since few agencies can readily bring such expertise together unless they frequently make different cross-sectional studies, there is considerable demand for the resources of firms providing market research and attitudinal survey expertise. Longitudinal studies done in 'rounds' (e.g. observation at, say, five-year intervals of historical time) can suffer in the same way but, with continuous monitoring, the problem disappears.

Any cross-sectional type of analysis can also be made with a longitudinal study but, in addition, the aggregation of changes in individuals can be looked at. This may make the additional expense worth while, though after the initial O_1, the sample becomes less and less representative of the population as individuals are selectively lost through death, migration, etc. Quite apart from the availability of incidence information, a longitudinal study makes it possible to see if certain events regularly precede others, e.g. 'Do people with a certain laboratory report consistently develop liver disease?' Knowledge about this time-sequencing may help in cause–effect arguments.

To be set against the longitudinal study is the greater cost although, of course, the basic need to define population and sample is common to both. At the report stage it may prove difficult to summarize the random sample of *histories* in a useful way. The problem of consistency of measurement techniques over a long period of time will have to be faced and may not have a satisfactory solution. Even with the most careful preliminary organization, unavoidable disintegration of the original project team through retiral, migration, etc., usually occurs to some extent. This may bring about degeneration of standards either through communication failures or because succeeding team members lack the pioneering zeal of the original staff. Frequent interim reports may help to maintain interest.

The overall length of a longitudinal study may be shortened somewhat by making it semi-longitudinal. In the typical case, a series of age cohorts is followed until the cohorts overlap. On the other hand, many studies become open-ended because more and more uses are found for a cohort whose past history is well documented. The National Child Development Study involving all 17 000 children born in Great Britain during the first week of March, 1958 is an example; although the initial interest centred on the effects of gestational experience on childhood development the cohort is still intermittently used in a variety of studies.

Another variant of the basic longitudinal study is exemplified by the British Labour Force Survey. Interviews are conducted throughout the year at the rate of 5000 households per month. Each sampled household is surveyed five times at quarterly intervals so that, every month, 1000 new households enter the study and 1000 households leave it. Thus, successive quarterly samples (of 15 000 households) have 12 000 households in common and the sampled households in two quarters one year apart have 3000 households in common.

18.4 INTERRUPTED TIME SERIES STUDIES

Many data are collected regularly and routinely, particularly by government agencies, local authorities, etc., so that, again, we have

$$\cdots O_1 \; O_2 \; O_3 \; O_4 \; O_5 \; O_6 \; O_7 \cdots$$

These observations are best thought of as a time series rather than a longitudinal study since the whole community of a country or region (rather than a sample) is generally involved. If, at some point, a legal reform, a 'crackdown' or a relatively abrupt social change, X, takes place, we have

$$\cdots O_1\ O_2\ O_3\ O_4\ X\ O_5\ O_6\ O_7\cdots$$

It may be possible to use this pattern – sometimes called an **interrupted time series** – to measure the effect of X, if any.

Sometimes a 'control' is available as when, for example, a change is made in England and Wales but not Scotland or when laws diverge in adjacent states of the USA. This allows estimation of history, etc., effects; but, even without such a control, the existence of observations extending on either side of X may permit elimination of some of the problems inherent in the simple $O_1 \rightarrow X \rightarrow O_2$ design of Chapter 15. If nothing else, we have a measure of the natural variability, but de-trending and de-seasonalization as discussed in Chapter 12 are also possible. The problem of subjects maturing seldom applies since we are usually looking at whole populations. Reactivity, learning and the O_1X interaction are eliminated because of the routineness of the O_i.

By considering a few examples we can isolate some of the problems that remain:

(a) Orlando Wilson

One well-known paradox relates to relative stability in the number of reported thefts in Chicago during the 20 years prior to the appointment (X) of one Orlando Wilson as police chief, followed by an abrupt increase and continued climb immediately after X, suggesting a considerable growth in crime as a result of Wilson's appointment. In reality, Wilson had tightened up reporting procedures so that fewer thefts escaped the record due to carelessness, corruption, etc. If we look at the corresponding figures for murders (which were much less likely to have been concealed), there is no sign of any change due to X. This is an example of an artefact due to **change in instrumentation** (more complete recording, in this case). It is always liable to be found when a 'new broom' arrives.

(b) Connecticut speeding crackdown

Another well-known example from the USA arises from an apparently striking fall in fatalities due to road accidents from one year to the next following a police drive against speeding motorists in Connecticut. However, when these two years' figures are compared with those of previous years, it becomes obvious that the crackdown was introduced *as a result* of a particularly bad year for accidents. We would therefore expect the following year(s) to be nearer the average. We have an example of selection bias as described in Section 15.2.7. The bad year in question could have had unusually severe winter weather but, even without information on this, figures from adjacent (and therefore similarly affected) states would provide controls.

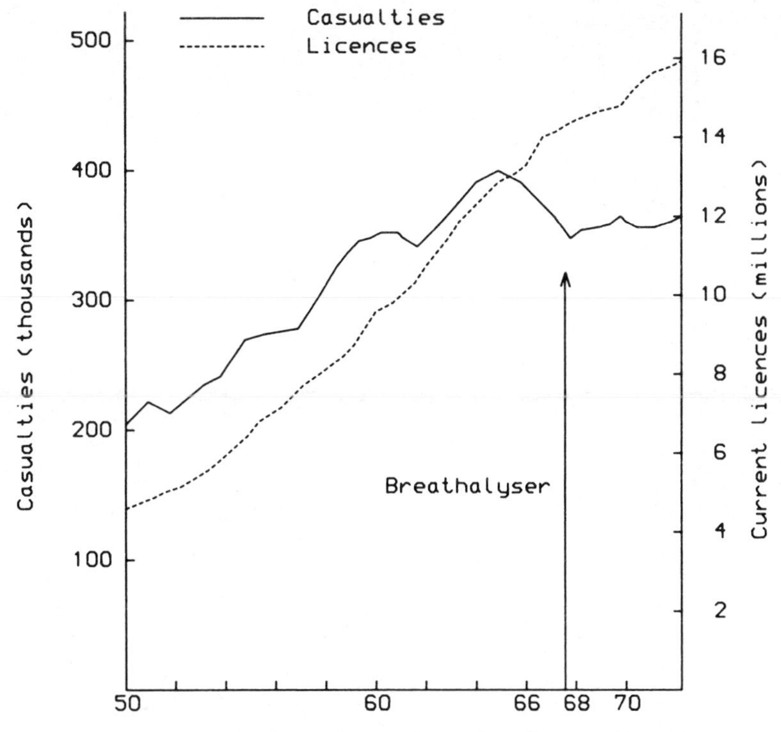

Fig. 18.1 Great Britain road casualties and car licences 1950–72

(c) The 'breathalyser'

A similar sort of crackdown in the UK involved introduction of the breathalyser in (September) 1967. Figure 18.1 shows how the total casualties fell over this period even though the number of vehicles on the road was increasing. But the fall began almost at the beginning of 1966! This was almost certainly due to drivers' increased awareness of the dangers of drunken driving resulting from publicity as the enabling act went through parliament. Of course, such prior advertisement of a new law is usually necessary for it to be an effectual deterrent. Were there drivers who changed their habits only through fear of personal detection? We can get some sort of control reference by comparing figures for fatalities and serious injuries (all of which would be reported) during commuting hours (when drivers are most likely to be sober) with those during weekend nights. Such a comparison leaves the effect of the breathalyser in little doubt.

(d) Compulsory wearing of car seat belts

Figure 18.2 is reproduced from *Statistical Bulletin* No. 2 (1985) of the Scottish Development Department and shows various death and injury patterns in

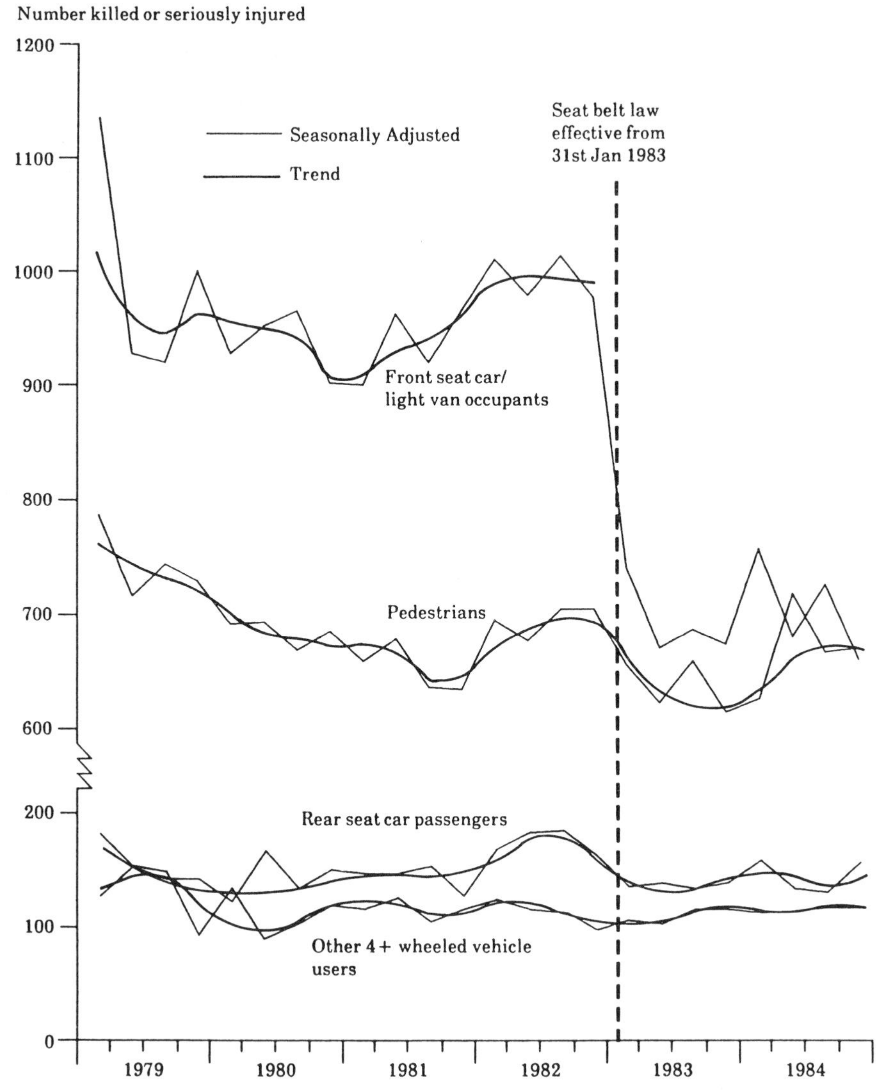

Fig. 18.2 Road accidents in Scotland: deaths and serious injuries, seaonally adjusted, February 1979–January 1985

Scotland spanning the introduction of compulsory wearing of seat belts by front-seat occupants of cars and light vans. The benefit of the legislation for these front-seat occupants is obvious. If drivers were simply being more careful, there would be a corresponding decline for pedestrians and rear seat passengers; a fall is indeed to be seen but is not nearly so marked. Indeed, it had actually been argued that the feeling of security induced in drivers by the wearing of a seat belt would increase the danger to other categories of road user; this is clearly not so.

(e) Whooping cough vaccination

Whooping cough vaccine is known to be effective in preventing the illness (or at least lessening its effects) in an individual. But it occasionally causes death or brain damage. It is therefore of some interest to ascertain whether the introduction of the vaccine in the UK round about 1957 was of benefit to the community as a whole. Figure 18.3 shows the notifications and deaths for the diagnosis since 1940. Notifications give a poor indication of incidence – witness the 'increase' when the National Health Service was introduced in 1947 and when notification became compulsory. Deaths, on the other hand (apart from epidemic

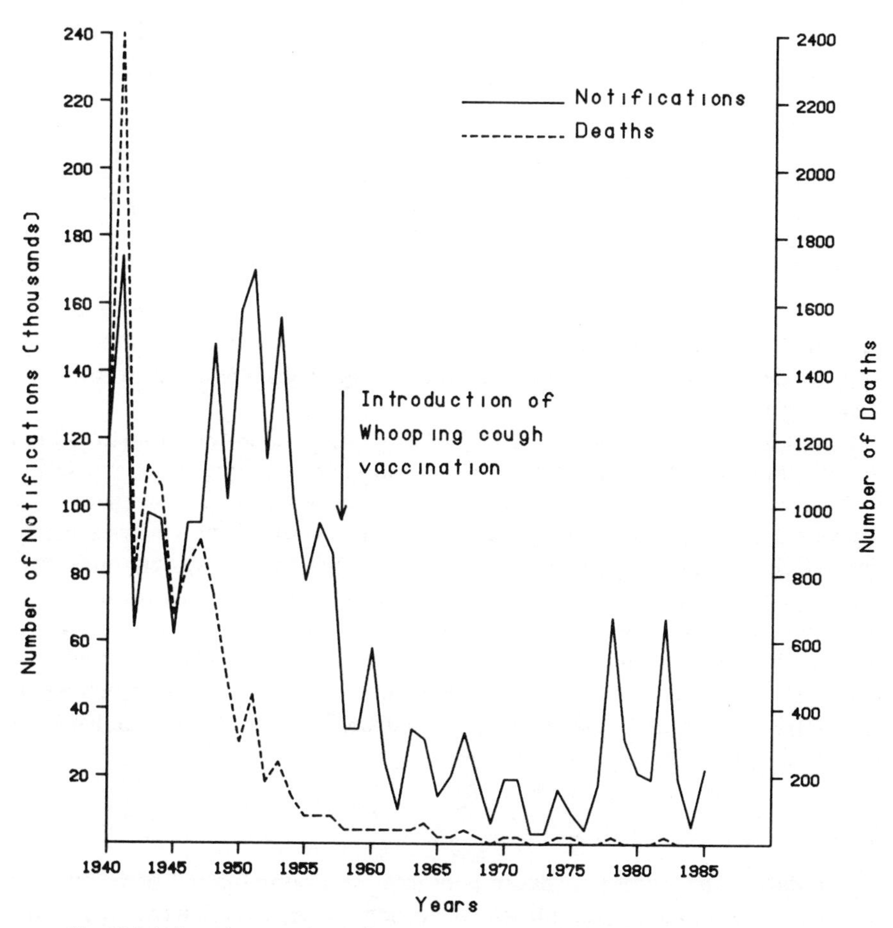

Fig. 18.3 Whooping cough notifications and deaths, England and Wales 1940–.

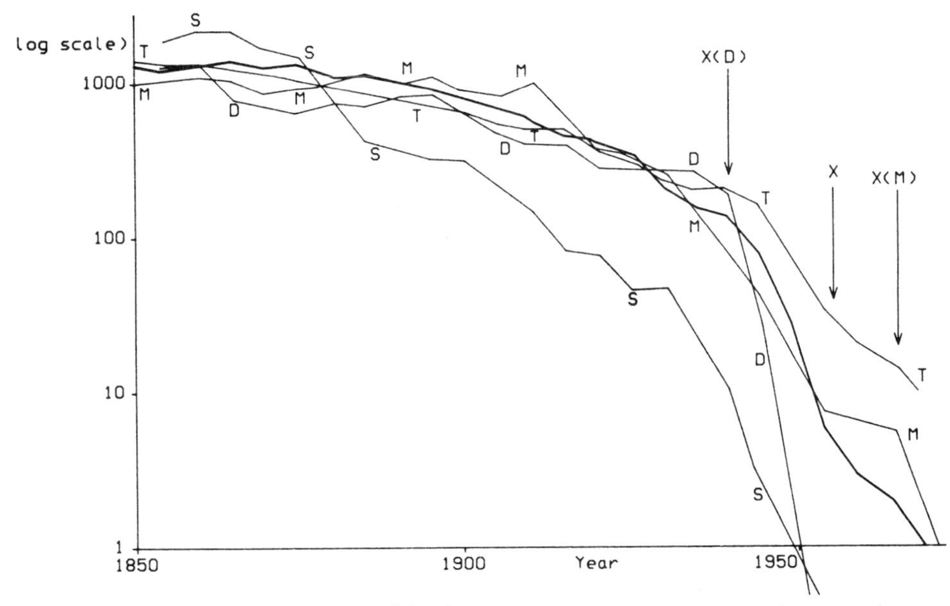

Fig. 18.4 Five-year moving average of death rates since 1850. × represents the start of an immunization programme Heavy line = whooping cough; D = diptheria; M = measles; S = scarlet fever; T = tuberculosis. For tuberculosis; standard mortality rates for all ages are given; for other diseases points represent death rates per million (ages < 15).

years such as 1941), show a falling trend due to better treatment of the illness, with little evidence of any considerable life-saving effect resulting from mass vaccination.

Figure 18.4 shows, by means of five-year moving averages of deaths per million in England and Wales, how the disease has behaved over a long period in relation to other serious illnesses. Tuberculosis, for which no vaccine is available, has shown a decline similar to whooping cough, as has scarlet fever which seemed to 'burn itself out'. Diphtheria and measles, against which mass vaccinations have also been instituted, showed fairly abrupt falls immediately after introduction of the respective vaccines and indicate what one might expect of a successful programme. One reason for the apparent failure of whooping cough immunization may lie in the ability of the virus to mutate to new forms which evade destruction until a new vaccine can be devised and introduced; this problem does not arise with diphtheria and measles.

SUMMARY

If incidence rather than prevalence is of interest, a longitudinal panel study is essential even though the administrative burden can be heavy. On the other hand, official statistics in the form of an interrupted time series present one of the simplest forms of investigation of social change, provided care is exercised in interpretation.

EXERCISES

1. Students' attitudes on university funding are to be studied by means of the following design:

$$X \to S_1 T_1$$
$$S_2 T_1 \to X' \to S_2' T_2$$

   where S_1 and S_2 are random samples of students starting their second and first years respectively interviewed at time T_1. Those S_2 students who survive into second year (i.e. S_2') are interviewed one year later at T_2. Thus, X and X' represent 'one year's experience of university life'.

   Discuss the design. Could it be made longitudinal?

2. Investigate the extent to which capital punishment is a deterrent to murder. Explain what statistical evidence you might examine (the experience of New Zealand is interesting) and how you would use the figures to come to as clear a conclusion as possible.

19
Smoking and lung cancer

19.1 INTRODUCTION

The object of this chapter is to show how the ideas presented in the earlier chapters can be used to collect and collate information on a specific question: 'Can smoking *cause* lung cancer?' Right from the introduction of the smoking habit to Europe, there have been doubts about the effect on health; only since the 1920s has evidence been gathered in any purposeful way.

19.2 EXPERIMENTS

In a case such as this, involving humans, it is not possible to perform a properly designed cause-to-effect experiment. It would be necessary to allocate at random a group of people to smoke (say) 50 cigarettes per day and another group to be non-smoker controls. But

1. there are obvious ethical problems
2. the experiment would need to run for very many years to demonstrate any difference.

However, animal experiments have shown that, if tobacco extracts are painted on rabbits' ears, cancerous growths tend to appear on the skin of the ear. On the other hand, beagles, taught with difficulty to 'smoke', appeared to suffer no ill effects to the lungs.

19.3 PUBLISHED FIGURES

Death certificates are not particularly useful since, as we have noted, the reported cause of death may be somewhat unreliable; in any case, no indication is given of the deceased's smoking habits. It is possible to compare lung cancer death rates across countries, sexes, classes, etc., where there are known to be smoking differences.

For instance, the Scottish death rate from lung cancer is at least twice that in countries like Eire and Iceland with a similar genetic background but lower tobacco consumption. But, of course, the Irish and the Icelanders tend to live in smaller, cleaner towns. The rate is increasing with time (1 in 900 in the UK in 1903 against 1 in 400 in 1973); but this may be an artefact of more accurate recording of the cause of death – before the arrival of modern terminal care, people usually died of pneumonia. Likewise the excess rate for males over females, for town dwellers over country dwellers, for social class V over social class I, all of which coincide with smoking differences, can be explained by other means.

'Odd' statistics may provide useful clues. For example, cancer of the hard palate is found worldwide but is fairly rare; however, the prevalence is very high in a certain area of South America. The explanation appears to be that shepherds at high altitudes reverse their cigars in their mouths to prevent them blowing out in the wind!

19.4 RETROSPECTIVE STUDIES (effect-to-cause)

In the 1920s and 1930s, doctors in the UK and USA began to realize that a very high proportion of their lung cancer patients had been heavy smokers. Because these comparisons were only among patients, the external validity was poor since patients are not a random sample of the general population.

Furthermore, we have only the patient's word for what he or she smoked; such estimates are often wildly 'out' and can be affected by what the patient thinks the doctor wants to hear.

Some tobacco companies even suggested that people might become smokers to relieve the irritation caused by a developing cancer! This possibility of precedence confusion can be dealt with as follows. There is a rare form of lung cancer, distinguishable from the usual type only under the microscope. It is found that people with the rare form are no more likely to be smokers than are those with no lung cancer. Since the irritation theory would apply equally to both types of cancer, the argument is refuted.

19.5 PROSPECTIVE STUDIES (cause-to-effect)

Raymond Pearl, a Baltimore general practitioner, followed the health records of hundreds of his white male patients for many years and, in 1938, published life tables for smokers ('moderate' and 'heavy') and non-smokers. The survivor curves are shown in Fig. 19.1

This suggested effects far beyond what could be accounted for by lung cancer deaths alone. But Pearl's study was criticized for the non-randomness of his sample.

It might be thought that one could look at lung cancer death rates among groups of people (e.g. Mormons, Seventh Day Adventists) who never smoke. But such people are abstemious in many other ways, e.g. they do not drink tea. Furthermore, suppose tendency to want to smoke and tendency to lung cancer are linked by inheritance; a smoking craving might cause some to leave the Mormon faith, but these are also the people with a tendency to lung cancer so that those left behind show a low incidence.

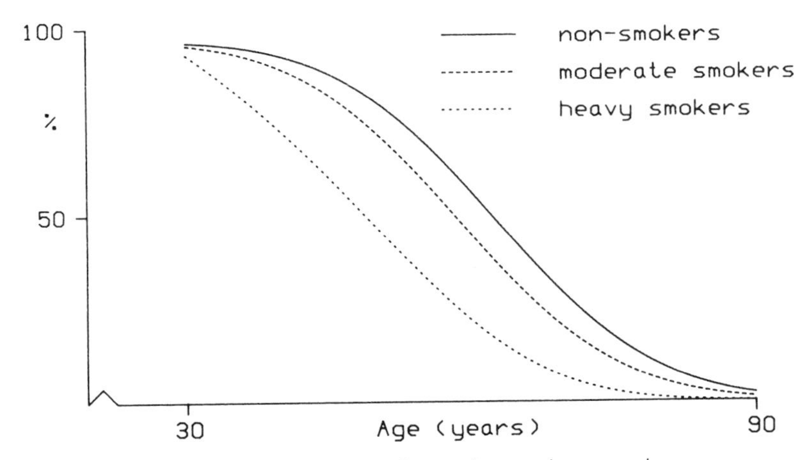

Fig. 19.1 Survivor curves for smokers and non-smokers

In 1951, two medical statisticians (Doll and Bradford-Hill) sent questionnaires to all doctors in the UK and received 40 000 replies (a two-thirds response rate) giving details of age, eating and exercise habits and, almost incidentally, smoking habits. (Because this was a prospective study there is no apparent reason why doctors would have lied about these.) With the aid of the Registrars General, the survival of these doctors was monitored for ten years and causes of all deaths determined. The results showed that only about 1 in 1000 non-smoker doctors died of lung cancer, whereas, for heavy smokers, the figure was 1 in 8. (The study also showed that the death rate from heart attacks was about 50% higher for smokers; this may help to explain Pearl's high death rates.)

At the time of the study, R.A. Fisher complained that there remained a logical loophole; the association could be the result of, say, a common genetic tendency. If this were so, giving up smoking would not alter an individual's chance of developing lung cancer. However, as a result of the Doll/Bradford-Hill report, many doctors did give up smoking and since 1960, the rate of death due to lung cancer has declined among doctors. This refutes Fisher's argument.

SUMMARY

In many statistical investigations, information from a variety of sources must be sifted and collated to produce a consistent and logical explanation of some phenomenon. The relationship between smoking and lung cancer is a classic study from the field of epidemiology.

EXERCISES

1. It is said that vegetarians (who do not eat meat or fish) are healthier and live longer than

meat-eaters. Discuss the difficulties involved in investigating these claims scientifically. How would you attempt to gather reliable evidence on the matter?

2. 'Heart disease' arises from the build-up of fat on the artery walls. The resulting blood pressure increase can rupture a blood vessel (haemorrhage) or a globule of fat can become detached and block a vessel (thrombosis). If either of these events happens, the result is a stroke or a heart attack according as the event is in the blood system of the brain or the heart muscle. Both can be severely disabling or fatal.
   The following causes have been suggested:
   (a) stress
   (b) soft water
   (c) lack of exercise
   (d) heredity
   (e) fat in the diet.
   Explain how you might seek evidence to evaluate these hypotheses.

3. Search out information about and write an essay on one of the following:
   (a) damage to the urban and rural environment due to 'acid rain',
   (b) clusters of cancer cases near nuclear power stations,
   (c) migration as a levelling influence on regional unemployment differentials.

20
An overview

In the original sense of the word, an 'interpreter' is one who negotiates, who is a broker between two parties. This book is therefore about the *two-way* process of seeking and assimilating the information that the world around is prepared to give – the data. We opened with the simile of a wobbling jelly; another parallel that may be helpful is the relationship between a shadow and its origin. The data are the shadow and, though description of that shadow can be of interest, it is the understanding of the source – the state of nature – that is the object of statistical analysis. To identify a fluttering shadow as the outline of a bird is not so revealing as the recognition that it is actually cast by a butterfly. This process can be assisted by trying to change the positions of the light source and the surface on which the shadow falls and then observing the changed pattern.

Although the reader may be entertained by debating the correctness of this analogy, its inclusion is intended to underline the principle that the design of a study and the analysis of the resulting information are intimately related. The chapters of this book switch erratically between these two aspects but they must come to blend in readers' minds so that their practical implications control the course of an investigation.

The aims of any programme of research should be fully specified at the outset. Often they are not. Such are the weaknesses of human nature that initial enthusiasm sometimes overcomes care and prudence in a desire to get the show on the road. Also, the task can be quite daunting; assumptions and approximations may be required that the originator of the project cannot accept, leading to revision of the goals or abandonment of the study. Just as students are often unwilling to check arithmetic because they would 'rather not know about any errors', so researchers are reluctant to examine their aims, hoping rather that all will become clear in the writing of the final report. If they are very lucky, it will. But the time to plan the final report is right at the start.

Of course, such insistence on exploring the logical steps of design and analysis may seem a gospel of perfection. Nevertheless, critical evaluation must be attempted and, if difficulties remain, they should at least be identified and their likely effects recognized and, if possible, quantified. Naturally, professional statisticians have considerable skill and experience on how to go about this and enjoy the challenge of participation, particularly if help is sought from an early stage.

In the application of its methods, statistics is both an art and a science. But the roots of the subject itself are in the realm of cognitive philosophy. Modern developments in artificial intelligence are bringing a new light to past perceptions of how the human mind interacts with nature. Certainly the most difficult ideas in statistics have little to do with mathematics but are rather about what is knowable and how it may be known. Although this may seem somewhat esoteric it is the first step towards understanding the concepts of population and sample with which this text began.

Glossary of symbols

The references in brackets refer to the main chapter and section in which the symbol is introduced, defined or used.

GENERAL

$\equiv$	is equivalent to
$\therefore$	therefore
‰	per thousand (10.1.1)

ROMAN SYMBOLS

B	Number of births (9.2)
$\mathrm{cov}(x, y)$	Covariance of quantitative variates x and y (4.2)
D	Number of deaths (9.2)
d_i	Standard deviation in stratum i (17.4.1)
d_x	Deaths at age x (10.1.6)
E	Number of emigrants (9.2)
e_x	Expectation of life at age x (10.1.6)
f	Observed frequency or proportion (3.5.2)
g	See Exercise 5 of Chapter 5
I	Number of immigrants (9.2)
L	Laspeyres price (quantity) index (8.4)
l_x	Numbers surviving to age x (10.1.6)
m	Death rate (10.1.2, 10.1.4 and 10.1.5)
N	Number in population (1.3)
N_i	Number in population in stratum i (17.4.1)
n	Number in sample (1.3)
n_1, n_2	Numbers in samples 1 and 2 (3.6)
n_i	Number in sample in stratum i (17.4.1)
O_i	Observations in a time series (18.2)
P	Paasche price (quantity) index (8.4)
P_a, P_b	Numbers (in age classes) in populations a and b (10.1.2)
P_t	Number in a population in year t (9.2)
p	Probability of the category of interest for the binomial distribution (5.3)
p_B, p_C	Base and current prices (8.2)

q	Probability of death (10.1.5 and 10.1.6)
q_B, q_C	Base and current quantities (8.2)
R	Random component of time series (12.3)
$R\{$	Randomization (15.4)
r	Count (discrete variate) (5.3)
r_{xy}	Correlation between continuous variates x and y (4.2)
S	Seasonal component of time series (12.3)
S_1, S_2	Two different samples (15.2.9)
S_{xx}	Corrected sum of squares for a continuous variate, x (3.4)
S_{xy}	Corrected sum of products between continuous variates, x and y (4.2)
s^2	Sample variance (3.4)
s_1^2, s_2^2	Sample variances in samples 1 and 2 (3.6)
s_p^2	Pooled sample variance (3.6)
s	Sample standard deviation (3.4)
T	Trend component of time series (12.3)
T_1, T_2	Two different time points (15.2)
T_x	See section 10.1.6
u_k	A random value equally likely to be anywhere between 0 and 1 (17.3)
v	Class mid-point (3.5.2)
w_i	weight (12.2.2)
X	An intervention or treatment (15.1)
x	Any continuous variate (3.4)
x_0	A specified value of x (4.3.1)
$\bar{x}$ ('x-bar')	The mean of x (3.4)
Y	Observed variable in a time series (12.3)
y	A continuous variate (4.1)
$\hat{y}$ ('y-hat')	Predicted value of y (4.3)
y_0	Observed value of y when $x = x_0$ (Exercise 3 of Chapter 5)
$\hat{y}_0$	Predicted value of y when $x = x_0$ (4.3.1)
y_t	Observed value of y at time t (12.2.1)
z	A continuous variate (4.4)

GREEK SYMBOLS

α (alpha)	Regression intercept (4.3)
$\hat{\alpha}$ ('alpha-hat')	Estimated regression intercept (4.3)
β (beta)	Regression slope (4.3)
$\hat{\beta}$ ('beta-hat')	Estimated regression slope (4.3)
ϕ (phi)	Sampling fraction (17.1)
ρ_{xy} (rho)	Population correlation between continuous variates x and y (15.2.7)
Σ (sigma)	Summation symbol (3.1.4)

Index

Adjective checklist 167
Age-heaping 90
Age–sex profile 113–21
Age-shifting 90
Age-specific death rate 123–5
Age-specific fertility rate 131–3
Alternative hypothesis 78
Area sampling 196
Arithmetic mean *see* Mean
Artificial intelligence 159, 215
Association 46–64, 66
Assumption 45, 73, 214
Average *see* Mean

Backing store 152
Bar diagram 12
Base point 103
BASIC 153
Batch working 153
Bias 3, 90, 163–73, 175, 192, 201
Bimodal distribution 34
Binomial distribution 68–76, 83
Birth 87
Birth cohort 119
Birth rate 131–3
Bivariate data 46–64
Blocking 184–8
British Labour Force Survey 203

Calibration 51
Categorical variate *see* Qualitative variate
Cause and effect 46, 59
Cause-to-effect 179, 211
CBR *see* Crude birth rate
CDR *see* Crude death rate
Censored observation 31
Census 1, 9, 86, 93, 165
Central processing unit 152
Centred moving average 146
CFM *see* Correction for mean
Chain-linked index number 107
Checklist 166
Class 20, 38
Class mid-point 15, 18
Class width 15, 16, *see also* Open-ended class
Clinical trial 188–9
Closed question 166
Cluster sampling 196
CMR *see* Crude marriage rate
Coding 91–2, 157, 161, 170–1
Cohort 119, 126, 129, 203
Computer 4, 53, 59, 71, 79, 86, 152–60, 193
Concordance 183
Conditional probability 66
Confidence interval 80
Confidentiality 165
Confounding 186–8
Consistency 79, 91, 98
Constant *see* Intercept
Continuous population register 88
Continuous variate 10, 20, 37
Control 178–81, 204, 205
Conversational mode 153
Corrected sum of squares 36, 37, 40, 157–8
Correction for mean 37
Correlation 48, 50, 54, 82
Cost 2, 46, 162, 164, 203
Covariance 47–8
CPU *see* Central processing unit
Cross-sectional study 201
Crude birth rate 131
Crude death rate 122–4, 130
Crude marriage rate 130
Cumulative frequency 18, 38
Current life table 129
Current position 103

Data collection 85–6, 154, 162–5
Data matrix 156
Data preparation 154–6
Data Protection Act 89, 156
Death 87, 122
Death rate 122–7
Decile 33
De facto census 86, 93
Degrees of freedom 36, 64
De jure census 86, 93
Demographic effect 134
Demographic transition 112–13
Demography 111–43
Deviation 34, 36, 51
Direct access 156
Discrete variate 10
Dispersion 36–7
Double-barrelled question 170
Double blind trial 180, 188
Double negative question 170

Edge effect 3
Effective digits 95–6
Effect-to-cause 179, 211
Estimation 80
Ethics 188, 210
Evidence 73–84
Expectation of life 128
Experiment 1, 53, 71, 85, 174–90, 210
Exponential distribution 20
Exponentially weighted moving average 146–7
External validity 180, 185, 193
Extrapolation 55
Extreme value distribution 21

Factor 186–8
Factorial treatment combination 186–8
Family Expenditure Survey 87
Fertility 131–3
Field coding 166
Filter question 166
Fixed format 155
Forecasting 146
Formal interview 163–4, 198
Forms mode entry 155
FORTRAN 153
Free format 155
Frequency distribution 18–32
Frequency distribution control 180
Frequency polygon 18–32
Frequency table 11–31, 38–40
Funnel effect 165

General fertility rate 131
General Household Survey 88, 89
Generation 119
Graphical summary 12–32
Grid 167
Gross reproduction rate 133
Group *see* Class
Group-administered questionnaire 164
Grouped data 14–32, 38–40

Halo effect 167
Hardware 152
Hawthorne effect 175
High-level language 153
Histogram 14–32, 70
History 174
Hospital Inpatient Survey 88
Hypothesis 72–84

ICD *see* International Statistical Classification of Diseases
Incidence 201
Incomplete response *see* Non-response
Independence 63, 66, 144
Independent samples 41
Index number 103
Index of Industrial Production 108
Index of Retail Prices 107
Inference 3, 9, 49, 65, 72–84
Instrumentation 175, 204
Interaction 57, 61, 187–8
Intercept 49–55
Internal validity 180, 185
International Passenger Survey 88
International Standard Classification of Occupations 92
International Statistical Classification of Diseases 92
Interpretation 214
Interquartile range 37, 39, 40
Interrupted time series 203–9
ISCD *see* International Statistical Classification of Diseases

Laspeyres 104–10
Leading question 169

Learning effect 175
Leverage 61
Life expectancy 128
Life table 127–9
Linear interpolation 39
Linear model *see* Model
Linear regression *see* Regression
Line diagram 144
Liveware 153
Loaded question 168
Location 33, 37, 40
Lognormal distribution 20, 32
Longitudinal study 201–3

Main effect 187
Male excess mortality 123
Margin 11, 12
Mark sensing 154
Marriage 87, 130–1
Matching 179–80, *see also* Paired comparison
Maturation 174
Mean 34–45, 80, *see also* Weighted mean
Mean deviation 36
Median 33–45, 75, 77–8, 80
Memory failure 169
Migration 133–4, 142
Missing value 158
Mode 34–8
Model 53, 55, 61, 62, 147
Mortality 174, *see also* Crude death rate
Moving average 144–7, 208
Multiple regression 53
Multiple tests 79
Multistage sampling 195–8
Multivariate data 46, 156

National Child Development Study 203
Natural decrease 112
Natural increase 112
Natural variation 3, 9, *see also* Sampling variation
Negative skewness 21
Net migration 112
Net reproduction rate 133
Nominal significance level 79
Non-linear model 53
Non-parametric test 78
Non-response 3, 162, 198
Normal distribution 20, 36, 71
Null hypothesis 73–84

Occupation bias 169
Official statistics *see* Published statistics
Ogive 18–32
One-sided alternative 78
Open-ended class 18, 40
Open question 166
Operating system 153
Optical scanning 154
Optimal allocation 194–5
Ordinal variate 10
Outlier 53

Paasche 105–10, 126
Paired comparison 42, 75, 77, 174–80, 183–4
Panel 202
Partial confounding 186
Percentile 33
Period 146
Period life table 129
Permuted block randomization 189
Pie diagram 12
Pilot study 161–2, 170, 194
Placebo 188
Pooled sample variance 42
Population 1, 9, 20, 35, 36, 51, *see also* Demography
Population distribution 20–1, 67–71
Population pyramid 113–21
Positive skewness 21
Postal questionnaire 162–3
Practical importance 80
Precision control 179
Prediction 49–53, 82, 139, *see also* Forecasting; Projection
Prestige bias 169
Pre-test *see* Pilot study
Prevalence 201
Price relative 103
Primary sampling unit 195–8
Probability 65–72, 201
Probability distribution 67, 70, 71
Probability proportional to size 196–8
Profile *see* Age–sex profile
Projection 139
Proportion 12, 33, 67, 83

Proportional allocation 194
Prospective study *see* Cause-to-effect
Published statistics 86–102, 210
Pyramid *see* Population pyramid

Quadrat 3, 8
Qualitative variate 9, 12, 46, 57
Quantitative variate 10, 14, 37, 46, 57
Quantity relative 103
Quartile 33
Questionnaire 2, 162–70
Question sequence 165
Quota sampling 198

Radix 127
Randomization 180, 184, 185, 189
Randomized block experiment 185
Random number table 4
Random sample 4–8, *see also* Sampling
Random variable *see* Qualitative variate; Quantitative variate
Range 36
Ranking 167
Rating 167
Reactivity 175
Reductio ad absurdum 72
Reduction of data 9
Regression 49–55
Regression coefficient 49–55
Relative frequency 34, 65
Reliability 90, 98, *see also* Projection
Residual 51–5, 61, 64, 150
Residual mean square 51, 53, 64
Residual sum of squares 51, 53
Residual variation 147, *see also* Regression
Retail Price Index *see* Index of Retail Prices
Retrospective study *see* Effect-to-cause
RMS *see* Residual mean square
RPI *see* Index of Retail Prices
RSS *see* Residual sum of squares

Sample 2
Sample statistic 33, 46, 51, 67
Sampling 1–8, 191–200
Sampling fraction 191
Sampling frame 191
Sampling variation 3, 9, 65–84
Scatter diagram 47–53
Scattergram *see* Scatter diagram
SDR *see* Standardized death rate
SE *see* Standard error
Seasonal variation 88, 147–8, 203, 204
Secondary analysis 85
Secondary sampling unit 195–8
Secular drift 201
Selection bias 175
Self-administered questionnaire 164
Significance test *see* Hypothesis
Sign test 78
Simple linear regression *see* Regression
Simple random sampling *see* Random sample
Simulation 72, 158
Single blind trial 188
Skewness 21, 37, 39, 40, 75
Slope, 49, 51–5
Smoothing 144
SMR *see* Standardized mortality ratio
Social class 92, 98
Socioeconomic group 92
Software 152
SPSS 153
Standard deviation 36–45, 71, 194
Standard error 41–5, 53, 71, 82, 198
Standard Industrial Classification 91
Standardized death rate 124–5
Standardized mortality ratio 126
Stratified sample 193–5
Stratum 193–5
Student's t-distribution 71
Subjective choice 3
Sum of squared residuals *see* Residual sum of squares
Survey 85, 161–73
Survivor plot 20, 21, 31
Symmetry 21, 35, *see also* Skewness
Systematic sampling 191–2

Tail 35
Tally 10, 14
Test *see* Hypothesis
Time series 144–51, *see also* Interrupted time series
Total fertility rate 132
Trend 147
Triangle test 74
Twin studies 183
Two-sided alternative 78

Two-way frequency table 11

Unit 1
Univariate data 46
Universe of discourse 1

Variance 36–45, 49, 53, 64
Variate *see* Qualitative variate;
Quantitative variate
VDU *see* Video display unit
Video display unit 153
Vital registration 87, 93

Weighted mean 35, 42, 104, 108
Weighting *see* Weighted mean